Understanding Global Warming

Understanding Global Warming

Declan Hernandez

R CALLISTO REFERENCE

www.callistoreference.com

Callisto Reference,
118-35 Queens Blvd., Suite 400,
Forest Hills, NY 11375, USA

Visit us on the World Wide Web at:
www.callistoreference.com

ISBN: 978-1-64116-602-7 (Hardback)

Cataloging-in-Publication Data

Understanding global warming / Declan Hernandez.
   p. cm.
Includes bibliographical references and index.
ISBN 978-1-64116-602-7
1. Global warming. 2. Global temperature changes. 3. Greenhouse effect, Atmospheric.
I. Hernandez, Declan.

QC981.8.G56 U53 2022
363.738 74--dc23

Table of Contents

Preface

The gradual increase in the average temperature of the Earth is known as global warming. It also refers to the ongoing increase in global surface temperature which is happening primarily due to human activities. The foremost cause of global warming is the rising levels of greenhouse gas emissions. Some of the major greenhouse gases are carbon dioxide, nitrous oxide and methane. There are various adverse changes which are caused due to global warming such as extreme weather events, expansion of deserts, rising sea levels and irregular precipitation. Change in temperature is observed to be the most severe in the arctic region, which has subsequently led to the retreat of glaciers, sea ice and permafrost. The book aims to shed light on some of the unexplored aspects of global warming and the recent researches in this field. The topics included herein about this problem are of utmost significance and bound to provide incredible insights to readers. Those in search of information to further their knowledge will be greatly assisted by this book.

Given below is the chapter wise description of the book:

Chapter 1- The long term rise in the average temperature of the Earth is known as global warming. It is an important aspect of the ongoing climate change and primarily refers to the temperature increase as a result of human activities. This is an introductory chapter which will introduce briefly all the significant aspects of global warming.

Chapter 2- Climate change refers to the phenomena where changes in the Earth's climate system lead to new weather patterns that stay for a substantial period of time. It can be caused due to natural reasons as well as human activities. This chapter closely examines these key causes of climate change as well as the external force mechanisms to provide an extensive understanding of the subject.

Chapter 3- Climatic system refers to the system whose components interact and give rise to Earth's climate. The recurring cyclical oscillations within regional or global climate are known as climate cycles. The topics elaborated in this chapter will help in gaining a better perspective about climate systems as well as the different climate cycles such as hydrological cycle and sulfur cycle.

Chapter 4- The process through which radiation from a planet's atmosphere warms the planet's surface to a temperature above compared to what it would be otherwise is known as the greenhouse effect. This chapter closely examines the key concepts related to the greenhouse effect such as its causes and greenhouse gases to provide an extensive understanding of the subject.

Chapter 5- There are various ways to control global warming such as using alternate energy sources and by using geoengineering. The topics elaborated in this chapter will help in gaining a better perspective about the different ways to manage global warming as well as related concepts such as the cause of global warming and global cooling.

At the end, I would like to thank all those who dedicated their time and efforts for the successful completion of this book. I also wish to convey my gratitude towards my friends and family who supported me at every step.

Declan Hernandez

Chapter 1

Introduction to Global Warming

The long term rise in the average temperature of the Earth is known as global warming. It is an important aspect of the ongoing climate change and primarily refers to the temperature increase as a result of human activities. This is an introductory chapter which will introduce briefly all the significant aspects of global warming.

Global warming is the phenomenon of increasing average air temperatures near the surface of Earth over the past one to two centuries. Climate scientists have since the mid-20th century gathered detailed observations of various weather phenomena (such as temperatures, precipitation, and storms) and of related influences on climate (such as ocean currents and the atmosphere's chemical composition). These data indicate that Earth's climate has changed over almost every conceivable timescale since the beginning of geologic time and that the influence of human activities since at least the beginning of the Industrial Revolution has been deeply woven into the very fabric of climate change.

Giving voice to a growing conviction of most of the scientific community, the Intergovernmental Panel on Climate Change (IPCC) was formed in 1988 by the World Meteorological Organization (WMO) and the United Nations Environment Program (UNEP). In 2013 the IPCC reported that the interval between 1880 and 2012 saw an increase in global average surface temperature of approximately 0.9 °C (1.5 °F). The increase is closer to 1.1 °C (2.0 °F) when measured relative to the preindustrial (i.e., 1750–1800) mean temperature.

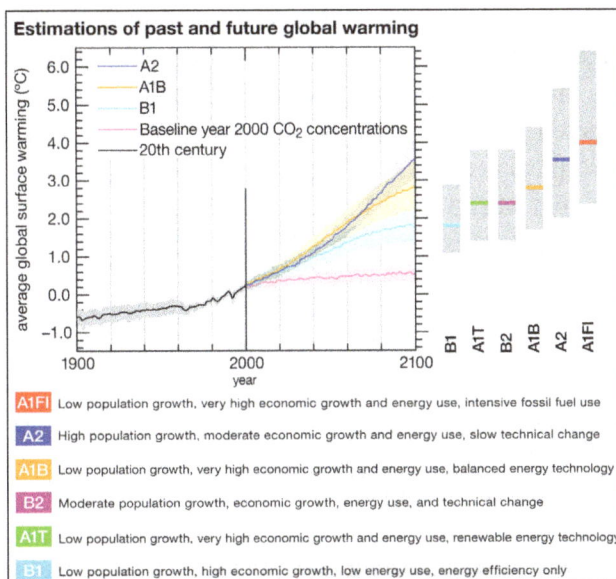

Estimations of past and future global warming

- A1FI Low population growth, very high economic growth and energy use, intensive fossil fuel use
- A2 High population growth, moderate economic growth and energy use, slow technical change
- A1B Low population growth, very high economic growth and energy use, balanced energy technology
- B2 Moderate population growth, economic growth, energy use, and technical change
- A1T Low population growth, very high economic growth and energy use, renewable energy technology
- B1 Low population growth, high economic growth, low energy use, energy efficiency only

Global warming scenarios.

Graph of the predicted increase in Earth's average surface temperature according to a series of climate change scenarios that assume different levels of economic development, population growth, and fossil fuel use. The assumptions made by each scenario are given at the bottom of the graph.

A special report produced by the IPCC in 2018 honed this estimate further, noting that human beings and human activities have been responsible for a worldwide average temperature increase of between 0.8 and 1.2 °C (1.4 and 2.2 °F) of global warming since preindustrial times, and most of the warming observed over the second half of the 20th century could be attributed to human activities. It predicted that the global mean surface temperature would increase between 3 and 4 °C (5.4 and 7.2 °F) by 2100 relative to the 1986–2005 average should carbon emissions continue at their current rate. The predicted rise in temperature was based on a range of possible scenarios that accounted for future greenhouse gas emissions and mitigation (severity reduction) measures and on uncertainties in the model projections. Some of the main uncertainties include the precise role of feedback processes and the impacts of industrial pollutants known as aerosols, which may offset some warming.

Many climate scientists agree that significant societal, economic, and ecological damage would result if global average temperatures rose by more than 2 °C (3.6 °F) in such a short time. Such damage would include increased extinction of many plant and animal species, shifts in patterns of agriculture, and rising sea levels. By 2015 all but a few national governments had begun the process of instituting carbon reduction plans as part of the Paris Agreement, a treaty designed to help countries keep global warming to 1.5 °C (2.7 °F) above preindustrial levels in order to avoid the worst of the predicted effects. Authors of a special report published by the IPCC in 2018 noted that should carbon emissions continue at their present rate, the increase in average near-surface air temperatures would reach 1.5 °C sometime between 2030 and 2052. Past IPCC assessments reported that the global average sea level rose by some 19–21 cm (7.5–8.3 inches) between 1901 and 2010 and that sea levels rose faster in the second half of the 20th century than in the first half. It also predicted, again depending on a wide range of scenarios, that the global average sea level would rise 26–77 cm (10.2–30.3 inches) relative to the 1986–2005 average by 2100 for global warming of 1.5 °C, an average of 10 cm (3.9 inches) less than what would be expected if warming rose to 2 °C (3.6 °F) above preindustrial levels.

The scenarios referred to above depend mainly on future concentrations of certain trace gases, called greenhouse gases, that have been injected into the lower atmosphere in increasing amounts through the burning of fossil fuels for industry, transportation, and residential uses. Modern global warming is the result of an increase in magnitude of the so-called greenhouse effect, a warming of Earth's surface and lower atmosphere caused by the presence of water vapour, carbon dioxide, methane, nitrous oxides, and other greenhouse gases. In 2014 the IPCC reported that concentrations of carbon dioxide, methane, and nitrous oxides in the atmosphere surpassed those found in ice cores dating back 800,000 years.

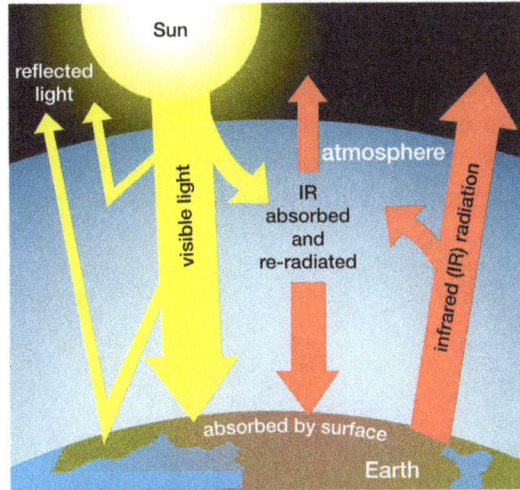

Greenhouse effect on Earth.

The greenhouse effect on Earth. Some incoming sunlight is reflected by Earth's atmosphere and surface, but most is absorbed by the surface, which is warmed. Infrared (IR) radiation is then emitted from the surface. Some IR radiation escapes to space, but some is absorbed by the atmosphere's greenhouse gases (especially water vapour, carbon dioxide, and methane) and reradiated in all directions, some to space and some back toward the surface, where it further warms the surface and the lower atmosphere.

Of all these gases, carbon dioxide is the most important, both for its role in the greenhouse effect and for its role in the human economy. It has been estimated that, at the beginning of the industrial age in the mid-18th century, carbon dioxide concentrations in the atmosphere were roughly 280 parts per million (ppm). By the middle of 2018 they had risen to 406 ppm, and, if fossil fuels continue to be burned at current rates, they are projected to reach 550 ppm by the mid-21st century—essentially, a doubling of carbon dioxide concentrations in 300 years.

A vigorous debate is in progress over the extent and seriousness of rising surface temperatures, the effects of past and future warming on human life, and the need for action to reduce future warming and deal with its consequences. It considers the causes of rising near-surface air temperatures, the influencing factors, the process of climate research and forecasting, the possible ecological and social impacts of rising temperatures, and the public policy developments since the mid-20th century.

Climatic Variation Since the Last Glaciation

Global warming is related to the more general phenomenon of climate change, which refers to changes in the totality of attributes that define climate. In addition to changes in air temperature, climate change involves changes to precipitation patterns, winds, ocean currents, and other measures of Earth's climate. Normally, climate change can be viewed as the combination of various natural forces occurring over diverse timescales. Since the advent of human civilization, climate change has involved an "anthropogenic," or exclusively human-caused, element, and this anthropogenic element has become more important in the industrial period of the past two centuries. The term global warming is used specifically to refer to any warming of near-surface air during the past two centuries that can be traced to anthropogenic causes.

A series of photographs of the Grinnell Glacier taken from the summit of Mount Gould in Glacier National Park.

To define the concepts of global warming and climate change properly, it is first necessary to recognize that the climate of Earth has varied across many timescales, ranging from an individual human life span to billions of years. This variable climate history is typically classified in terms of "regimes" or "epochs." For instance, the Pleistocene glacial epoch (about 2,600,000 to 11,700 years ago) was marked by substantial variations in the global extent of glaciers and ice sheets. These variations took place on timescales of tens to hundreds of millennia and were driven by changes in the distribution of solar radiation across Earth's surface. The distribution of solar radiation is known as the insolation pattern, and it is strongly affected by the geometry of Earth's orbit around the Sun and by the orientation, or tilt, of Earth's axis relative to the direct rays of the Sun.

Worldwide, the most recent glacial period, or ice age, culminated about 21,000 years ago in what is often called the Last Glacial Maximum. During this time, continental ice sheets extended well into the middle latitude regions of Europe and North America, reaching as far south as present-day London and New York City. Global annual mean temperature appears to have been about 4–5 °C (7–9 °F) colder than in the mid-20th century. It is important to remember that these figures are a global average. In fact, during the height of this last ice age, Earth's climate was characterized by greater cooling at higher latitudes (that is, toward the poles) and relatively little cooling over large parts of the tropical oceans (near the Equator). This glacial interval terminated abruptly about 11,700 years ago and was followed by the subsequent relatively ice-free period known as the Holocene Epoch. The modern period of Earth's history is conventionally defined as residing within the Holocene. However, some scientists have argued that the Holocene Epoch terminated in the relatively recent past and that Earth currently resides in a climatic interval that could justly be called the Anthropocene Epoch—that is, a period during which humans have exerted a dominant influence over climate.

Though less dramatic than the climate changes that occurred during the Pleistocene Epoch, significant variations in global climate have nonetheless taken place over the course of the Holocene. During the early Holocene, roughly 9,000 years ago, atmospheric circulation and precipitation patterns appear to have been substantially different from those of today. For example, there is evidence for relatively wet conditions in what is now the Sahara Desert. The change from one climatic regime to another was caused by only modest changes in the pattern of insolation within the Holocene interval as well as the interaction of these patterns with large-scale climate phenomena such as monsoons and El Niño/Southern Oscillation (ENSO).

During the middle Holocene, some 5,000–7,000 years ago, conditions appear to have been relatively warm—indeed, perhaps warmer than today in some parts of the world and during certain seasons. For this reason, this interval is sometimes referred to as the Mid-Holocene Climatic Optimum. The relative warmth of average near-surface air temperatures at this time, however, is somewhat unclear. Changes in the pattern of insolation favoured warmer summers at higher latitudes in the Northern Hemisphere, but these changes also produced cooler winters in the Northern Hemisphere and relatively cool conditions year-round in the tropics. Any overall hemispheric or global mean temperature changes thus reflected a balance between competing seasonal and regional changes. In fact, recent theoretical climate model studies suggest that global mean temperatures during the middle Holocene were probably 0.2–0.3 °C (0.4–0.5 °F) colder than average late 20th-century conditions.

Over subsequent millennia, conditions appear to have cooled relative to middle Holocene levels. This period has sometimes been referred to as the "Neoglacial." In the middle latitudes this cooling trend was associated with intermittent periods of advancing and retreating mountain glaciers reminiscent of (though far more modest than) the more substantial advance and retreat of the major continental ice sheets of the Pleistocene climate epoch.

Causes of Global Warming

Greenhouse Effect

The average surface temperature of Earth is maintained by a balance of various forms of solar and terrestrial radiation. Solar radiation is often called "shortwave" radiation because the frequencies of the radiation are relatively high and the wavelengths relatively short—close to the visible portion of the electromagnetic spectrum. Terrestrial radiation, on the other hand, is often called "longwave" radiation because the frequencies are relatively low and the wavelengths relatively long—somewhere in the infrared part of the spectrum. Downward-moving solar energy is typically measured in watts per square metre. The energy of the total incoming solar radiation at the top of Earth's atmosphere (the so-called "solar constant") amounts roughly to 1,366 watts per square metre annually. Adjusting for the fact that only one-half of the planet's surface receives solar radiation at any given time, the average surface insolation is 342 watts per square metre annually.

The amount of solar radiation absorbed by Earth's surface is only a small fraction of the total solar radiation entering the atmosphere. For every 100 units of incoming solar radiation, roughly 30 units are reflected back to space by either clouds, the atmosphere, or reflective regions of Earth's surface. This reflective capacity is referred to as Earth's planetary albedo, and it need not remain fixed over time, since the spatial extent and distribution of reflective formations, such as clouds and ice cover, can change. The 70 units of solar radiation that are not reflected may be absorbed by the atmosphere, clouds, or the surface. In the absence of further complications, in order to maintain thermodynamic equilibrium, Earth's surface and atmosphere must radiate these same 70 units back to space. Earth's surface temperature (and that of the lower layer of the atmosphere essentially in contact with the surface) is tied to the magnitude of this emission of outgoing radiation according to the Stefan-Boltzmann law.

Earth's energy budget is further complicated by the greenhouse effect. Trace gases with certain chemical properties—the so-called greenhouse gases, mainly carbon dioxide (CO_2), methane (CH_4), and nitrous oxide (N_2O)—absorb some of the infrared radiation produced by Earth's surface. Because of this absorption, some fraction of the original 70 units does not directly escape to space. Because greenhouse gases emit the same amount of radiation they absorb and because this radiation is emitted equally in all directions (that is, as much downward as upward), the net effect of absorption by greenhouse gases is to increase the total amount of radiation emitted downward toward Earth's surface and lower atmosphere. To maintain equilibrium, Earth's surface and lower atmosphere must emit more radiation than the original 70 units. Consequently, the surface temperature must be higher. This process is not quite the same as that which governs a true greenhouse, but the end effect is similar. The presence of greenhouse gases in the atmosphere leads to a warming of the surface and lower part of the atmosphere (and a cooling higher up in the atmosphere) relative to what would be expected in the absence of greenhouse gases.

It is essential to distinguish the "natural," or background, greenhouse effect from the "enhanced" greenhouse effect associated with human activity. The natural greenhouse effect is associated with surface warming properties of natural constituents of Earth's atmosphere, especially water vapour, carbon dioxide, and methane. The existence of this effect is accepted by all scientists. Indeed, in its absence, Earth's average temperature would be approximately 33 °C (59 °F) colder than today, and Earth would be a frozen and likely uninhabitable planet. What has been subject to controversy is the so-called enhanced greenhouse effect, which is associated with increased concentrations of greenhouse gases caused by human activity. In particular, the burning of fossil fuels raises the concentrations of the major greenhouse gases in the atmosphere, and these higher concentrations have the potential to warm the atmosphere by several degrees.

Radiative Forcing

The greenhouse effect, it is apparent that the temperature of Earth's surface and lower atmosphere may be modified in three ways: through a net increase in the solar radiation entering at the top of Earth's atmosphere, through a change in the fraction of the radiation reaching the surface, and through a change in the concentration of greenhouse gases in the atmosphere. In each case the changes can be thought of in terms of "radiative forcing." As defined by the IPCC, radiative forcing is a measure of the influence a given climatic factor has on the amount of downward-directed radiant energy impinging upon Earth's surface. Climatic factors are divided between those caused primarily by human activity (such as greenhouse gas emissions and aerosol emissions) and those caused by natural forces (such as solar irradiance); then, for each factor, so-called forcing values are calculated for the time period between 1750 and the present day. "Positive forcing" is exerted by climatic factors that contribute to the warming of Earth's surface, whereas "negative forcing" is exerted by factors that cool Earth's surface.

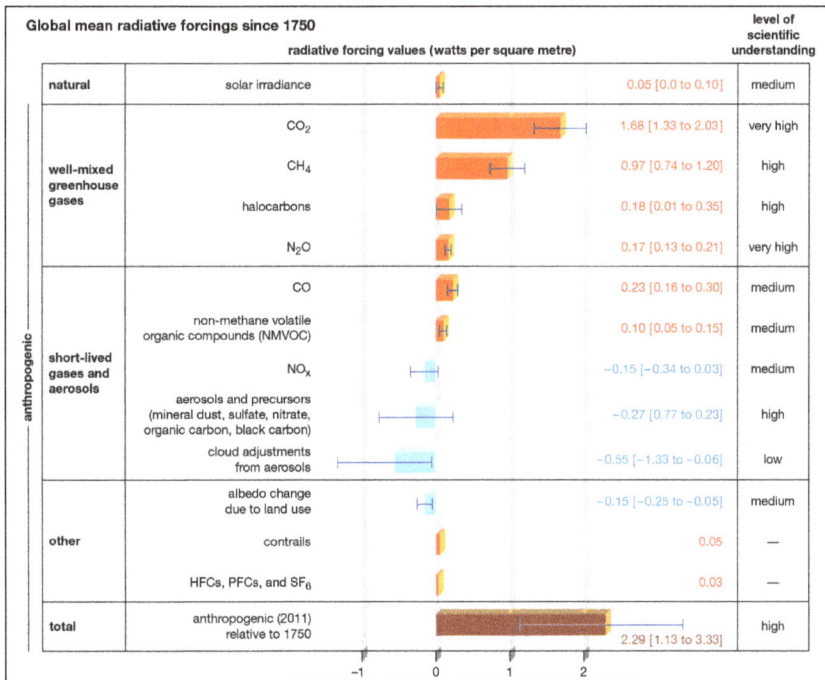

Global mean radiative forcings since 1750		radiative forcing values (watts per square metre)	level of scientific understanding
natural	solar irradiance	0.05 [0.0 to 0.10]	medium
well-mixed greenhouse gases	CO_2	1.68 [1.33 to 2.03]	very high
	CH_4	0.97 [0.74 to 1.20]	high
	halocarbons	0.18 [0.01 to 0.35]	high
	N_2O	0.17 [0.13 to 0.21]	very high
short-lived gases and aerosols	CO	0.23 [0.16 to 0.30]	medium
	non-methane volatile organic compounds (NMVOC)	0.10 [0.05 to 0.15]	medium
	NO_x	−0.15 [−0.34 to 0.03]	medium
	aerosols and precursors (mineral dust, sulfate, nitrate, organic carbon, black carbon)	−0.27 [0.77 to 0.23]	high
	cloud adjustments from aerosols	−0.55 [−1.33 to −0.06]	low
other	albedo change due to land use	−0.15 [−0.25 to −0.05]	medium
	contrails	0.05	—
	HFCs, PFCs, and SF_6	0.03	—
total	anthropogenic (2011) relative to 1750	2.29 [1.13 to 3.33]	high

The concentration of carbon dioxide and other greenhouse gases has increased in Earth's atmosphere. As a result of these and other factors, Earth's atmosphere retains more heat than in the past.

On average, about 342 watts of solar radiation strike each square metre of Earth's surface per year, and this quantity can in turn be related to a rise or fall in Earth's surface temperature. Temperatures at the surface may also rise or fall through a change in the distribution of terrestrial radiation (that is, radiation emitted by Earth) within the atmosphere. In some cases, radiative forcing has a natural origin, such as during explosive eruptions from volcanoes where vented gases and ash block some portion of solar radiation from the surface. In other cases, radiative forcing has an anthropogenic, or exclusively human, origin. For example, anthropogenic increases in carbon dioxide, methane, and nitrous oxide are estimated to account for 2.3 watts per square metre of positive radiative forcing. When all values of positive and negative radiative forcing are taken together and all interactions between climatic factors are accounted for, the total net increase in surface radiation due to human activities since the beginning of the Industrial Revolution is 1.6 watts per square metre.

Influences of Human Activity on Climate

Human activity has influenced global surface temperatures by changing the radiative balance governing the Earth on various timescales and at varying spatial scales. The most profound and well-known anthropogenic influence is the elevation of concentrations of greenhouse gases in the atmosphere. Humans also influence climate by changing the concentrations of aerosols and ozone and by modifying the land cover of Earth's surface.

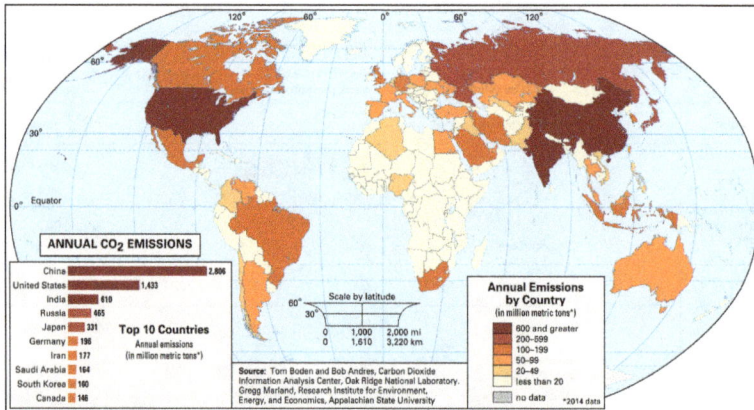

Carbon dioxide emissions.

Greenhouse Gases

Factories that burn fossil fuels help to cause global warming.

As discussed above, greenhouse gases warm Earth's surface by increasing the net downward longwave radiation reaching the surface. The relationship between atmospheric concentration of greenhouse gases and the associated positive radiative forcing of the surface is different for each gas. A complicated relationship exists between the chemical properties of each greenhouse gas and the relative amount of longwave radiation that each can absorb.

Water Vapour

Water vapour is the most potent of the greenhouse gases in Earth's atmosphere, but its behaviour is fundamentally different from that of the other greenhouse gases. The primary role of water vapour is not as a direct agent of radiative forcing but rather as a climate feedback—that is, as a response within the climate system that influences the system's continued activity. This distinction arises from the fact that the amount of water vapour in the atmosphere cannot, in general, be directly modified by human behaviour but is instead set by air temperatures. The warmer the surface, the greater the evaporation rate of water from the surface. As a result, increased evaporation leads to a greater concentration of water vapour in the lower atmosphere capable of absorbing longwave radiation and emitting it downward.

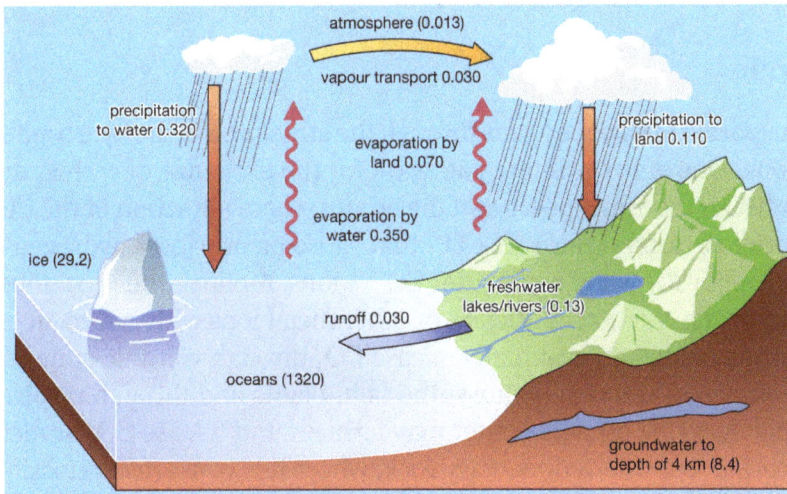

The present-day surface hydrologic cycle, in which water is transferred from the oceans through the atmosphere to the continents and back to the oceans over and beneath the land surface. The values in parentheses following the various forms of water (e.g. ice) refer to volumes in millions of cubic kilometres; those following the processes (e.g. precipitation) refer to their fluxes in millions of cubic kilometres of water per year.

Carbon Dioxide

Of the greenhouse gases, carbon dioxide (CO_2) is the most significant. Natural sources of atmospheric CO_2 include outgassing from volcanoes, the combustion and natural decay of organic matter, and respiration by aerobic (oxygen-using) organisms. These sources are balanced, on average, by a set of physical, chemical, or biological processes, called "sinks," that tend to remove CO_2 from the atmosphere. Significant natural sinks include terrestrial vegetation, which takes up CO_2 during the process of photosynthesis.

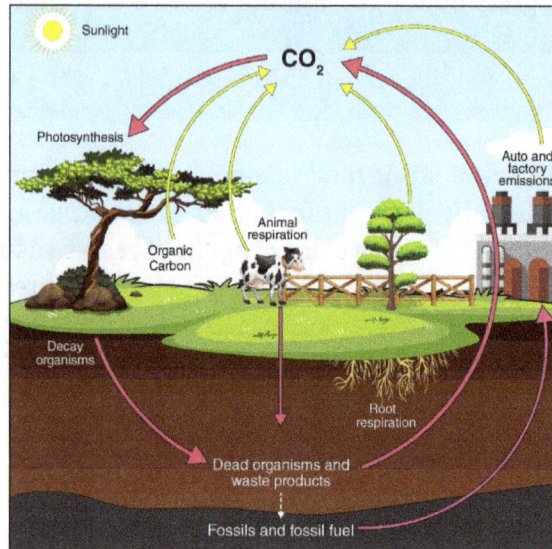

Carbon Cycle

Carbon is transported in various forms through the atmosphere, the hydrosphere, and geologic formations. One of the primary pathways for the exchange of carbon dioxide (CO_2) takes place between the atmosphere and the oceans; there a fraction of the CO_2 combines with water, forming carbonic acid (H_2CO_3) that subsequently loses hydrogen ions (H^+) to form bicarbonate (HCO_3^-) and carbonate (CO_3^{2-}) ions. Mollusk shells or mineral precipitates that form by the reaction of calcium or other metal ions with carbonate may become buried in geologic strata and eventually release CO_2 through volcanic outgassing. Carbon dioxide also exchanges through photosynthesis in plants and through respiration in animals. Dead and decaying organic matter may ferment and release CO_2 or methane (CH_4) or may be incorporated into sedimentary rock, where it is converted to fossil fuels. Burning of hydrocarbon fuels returns CO_2 and water (H_2O) to the atmosphere. The biological and anthropogenic pathways are much faster than the geochemical pathways and, consequently, have a greater impact on the composition and temperature of the atmosphere.

A number of oceanic processes also act as carbon sinks. One such process, called the "solubility pump," involves the descent of surface seawater containing dissolved CO_2. Another process, the "biological pump," involves the uptake of dissolved CO_2 by marine vegetation and phytoplankton (small free-floating photosynthetic organisms) living in the upper ocean or by other marine organisms that use CO_2 to build skeletons and other structures made of calcium carbonate ($CaCO_3$). As these organisms expire and fall to the ocean floor, the carbon they contain is transported downward and eventually buried at depth. A long-term balance between these natural sources and sinks leads to the background, or natural, level of CO_2 in the atmosphere.

In contrast, human activities increase atmospheric CO_2 levels primarily through the burning of fossil fuels—principally oil and coal and secondarily natural gas, for use in transportation, heating, and the generation of electrical power—and through the production of cement. Other anthropogenic sources include the burning of forests and the clearing of land. Anthropogenic emissions currently account for the annual release of about 7 gigatons (7 billion tons) of carbon into the atmosphere. Anthropogenic emissions are equal to approximately 3 percent of the total emissions of CO_2 by natural sources, and this amplified carbon load from human activities far exceeds the offsetting capacity of natural sinks (by perhaps as much as 2–3 gigatons per year).

Deforestation: Smoldering remains of a plot of deforested land in the Amazon Rainforest of Brazil. Annually, it is estimated that net global deforestation accounts for about two gigatons of carbon emissions to the atmosphere.

CO_2 consequently accumulated in the atmosphere at an average rate of 1.4 ppm per year between 1959 and 2006 and roughly 2.0 ppm per year between 2006 and 2018. Overall, this rate of accumulation has been linear (that is, uniform over time). However, certain current sinks, such as the oceans, could become sources in the future. This may lead to a situation in which the concentration of atmospheric CO_2 builds at an exponential rate (that is, its rate of increase is also increasing).

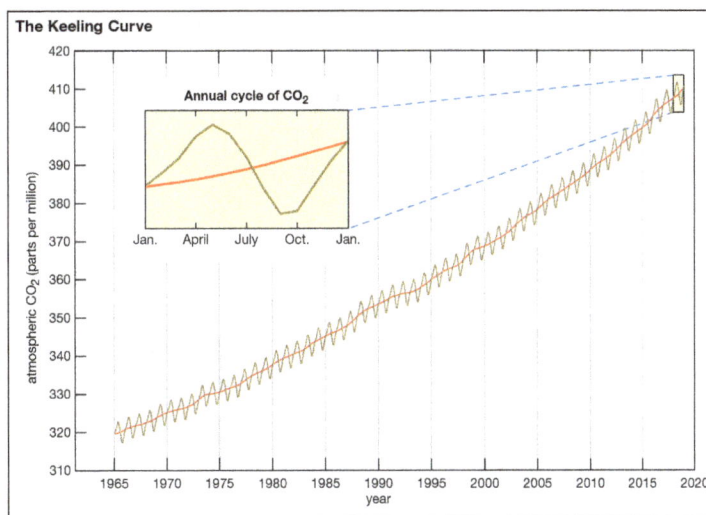

The Keeling Curve, named after American climate scientist Charles David Keeling, tracks changes in the concentration of carbon dioxide (CO_2) in Earth's atmosphere at a research station on Mauna Loa in Hawaii. Although these concentrations experience small seasonal fluctuations, the overall trend shows that CO_2 is increasing in the atmosphere.

The natural background level of carbon dioxide varies on timescales of millions of years because of slow changes in outgassing through volcanic activity. For example, roughly 100 million years ago, during the Cretaceous Period (145 million to 66 million years ago), CO_2 concentrations appear to have been several times higher than they are today (perhaps close to 2,000 ppm). Over the past 700,000 years, CO_2 concentrations have varied over a far smaller range (between roughly 180 and 300 ppm) in association with the same Earth orbital effects linked to the coming and going of the Pleistocene ice ages. By the early 21st century, CO_2 levels had reached 384 ppm, which is approximately 37 percent above the natural background level of roughly 280 ppm that existed at the beginning of the Industrial Revolution. Atmospheric CO_2 levels continued to increase, and by 2018 they had reached 410 ppm. Such levels are believed to be the highest in at least 800,000 years according to ice core measurements and may be the highest in at least 5 million years according to other lines of evidence.

Radiative forcing caused by carbon dioxide varies in an approximately logarithmic fashion with the concentration of that gas in the atmosphere. The logarithmic relationship occurs as the result of a saturation effect wherein it becomes increasingly difficult, as CO_2 concentrations increase, for additional CO_2 molecules to further influence the "infrared window" (a certain narrow band of wavelengths in the infrared region that is not absorbed by atmospheric gases). The logarithmic relationship predicts that the surface warming potential will rise by roughly the same amount for each doubling of CO_2 concentration. At current rates of fossil fuel use, a doubling of CO_2 concentrations over preindustrial levels is expected to take place by the middle of the 21st century (when CO_2 concentrations are projected to reach 560 ppm). A doubling of CO_2 concentrations would represent an increase of roughly 4 watts per square metre of radiative forcing. Given typical estimates of "climate sensitivity" in the absence of any offsetting factors, this energy increase would lead to a warming of 2 to 5 °C (3.6 to 9 °F) over preindustrial times. The total radiative forcing by anthropogenic CO_2 emissions since the beginning of the industrial age is approximately 1.66 watts per square metre.

Methane

Methane (CH_4) is the second most important greenhouse gas. CH_4 is more potent than CO_2 because the radiative forcing produced per molecule is greater. In addition, the infrared window is less saturated in the range of wavelengths of radiation absorbed by CH_4, so more molecules may fill in the region. However, CH_4 exists in far lower concentrations than CO_2 in the atmosphere, and its concentrations by volume in the atmosphere are generally measured in parts per billion (ppb) rather than ppm. CH_4 also has a considerably shorter residence time in the atmosphere than CO_2 (the residence time for CH_4 is roughly 10 years, compared with hundreds of years for CO_2).

Natural sources of methane include tropical and northern wetlands, methane-oxidizing bacteria that feed on organic material consumed by termites, volcanoes, seepage vents of the seafloor in regions rich with organic sediment, and methane hydrates trapped along the continental shelves of the oceans and in polar permafrost. The primary natural sink for methane is the atmosphere itself, as methane reacts readily with the hydroxyl radical ($\cdot$OH) within the troposphere to form CO_2 and water vapour (H_2O). When CH_4 reaches the stratosphere, it is destroyed. Another natural sink is soil, where methane is oxidized by bacteria.

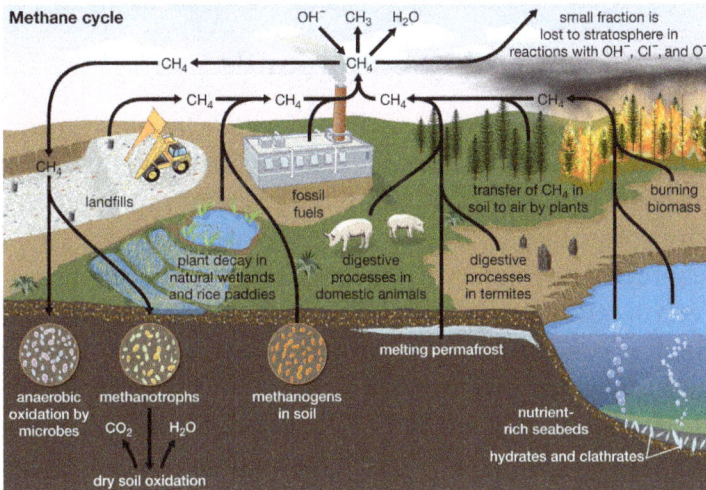

Methane cycle.

As with CO_2, human activity is increasing the CH_4 concentration faster than it can be offset by natural sinks. Anthropogenic sources currently account for approximately 70 percent of total annual emissions, leading to substantial increases in concentration over time. The major anthropogenic sources of atmospheric CH_4 are rice cultivation, livestock farming, the burning of coal and natural gas, the combustion of biomass, and the decomposition of organic matter in landfills. Future trends are particularly difficult to anticipate. This is in part due to an incomplete understanding of the climate feedbacks associated with CH_4 emissions. In addition it is difficult to predict how, as human populations grow, possible changes in livestock raising, rice cultivation, and energy utilization will influence CH_4 emissions.

It is believed that a sudden increase in the concentration of methane in the atmosphere was responsible for a warming event that raised average global temperatures by 4–8 °C (7.2–14.4 °F) over a few thousand years during the so-called Paleocene-Eocene Thermal Maximum, or PETM. This episode took place roughly 55 million years ago, and the rise in CH_4 appears to have been related to a massive volcanic eruption that interacted with methane-containing flood deposits. As a result, large amounts of gaseous CH_4 were injected into the atmosphere. It is difficult to know precisely how high these concentrations were or how long they persisted. At very high concentrations, residence times of CH_4 in the atmosphere can become much greater than the nominal 10-year residence time that applies today. Nevertheless, it is likely that these concentrations reached several ppm during the PETM.

Methane concentrations have also varied over a smaller range (between roughly 350 and 800 ppb) in association with the Pleistocene ice age cycles. Preindustrial levels of CH_4 in the atmosphere were approximately 700 ppb, whereas levels exceeded 1,867 ppb in late 2018. (These concentrations are well above the natural levels observed for at least the past 650,000 years.) The net radiative forcing by anthropogenic CH_4 emissions is approximately 0.5 watt per square metre—or roughly one-third the radiative forcing of CO_2.

Surface-level Ozone and other Compounds

The next most significant greenhouse gas is surface, or low-level, ozone (O_3). Surface O_3 is a result of air pollution; it must be distinguished from naturally occurring stratospheric O_3, which has a very different role in the planetary radiation balance. The primary natural source of surface O_3 is the subsidence of stratospheric O_3 from the upper atmosphere. In contrast, the primary anthropogenic source of surface O_3 is photochemical reactions involving the atmospheric pollutant carbon monoxide (CO). The best estimates of the natural concentration of surface O_3 are 10 ppb, and the net radiative forcing due to anthropogenic emissions of surface O_3 is approximately 0.35 watt per square metre. Ozone concentrations can rise above unhealthy levels (that is, conditions where concentrations meet or exceed 70 ppb for eight hours or longer) in cities prone to photochemical smog.

Nitrous Oxides and Fluorinated Gases

Additional trace gases produced by industrial activity that have greenhouse properties include nitrous oxide (N_2O) and fluorinated gases (halocarbons), the latter including sulfur hexafluoride, hydrofluorocarbons (HFCs), and perfluorocarbons (PFCs). Nitrous oxide is responsible for 0.16 watt per square metre radiative forcing, while fluorinated gases are collectively responsible for 0.34 watt per square metre. Nitrous oxides have small background concentrations due to natural biological reactions in soil and water, whereas the fluorinated gases owe their existence almost entirely to industrial sources.

Aerosols

The production of aerosols represents an important anthropogenic radiative forcing of climate. Collectively, aerosols block—that is, reflect and absorb—a portion of incoming solar radiation, and this creates a negative radiative forcing. Aerosols are second only to greenhouse gases in relative importance in their impact on near-surface air temperatures. Unlike the decade-long residence times of the "well-mixed" greenhouse gases, such as CO_2 and CH_4, aerosols are readily flushed out of the atmosphere within days, either by rain or snow (wet deposition) or by settling out of the air (dry deposition). They must therefore be continually generated in order to produce a steady effect on radiative forcing. Aerosols have the ability to influence climate directly by absorbing or reflecting incoming solar radiation, but they can also produce indirect effects on climate by modifying cloud formation or cloud properties. Most aerosols serve as condensation nuclei (surfaces upon which water vapour can condense to form clouds); however, darker-coloured aerosols may hinder cloud formation by absorbing sunlight and heating up the surrounding air. Aerosols can be transported thousands of kilometres from their sources of origin by winds and upper-level circulation in the atmosphere.

Perhaps the most important type of anthropogenic aerosol in radiative forcing is sulfate aerosol. It is produced from sulfur dioxide (SO_2) emissions associated with the burning of coal and oil. Since the late 1980s, global emissions of SO_2 have decreased from about 151.5 million tonnes (167.0 million tons) to less than 100 million tonnes (110.2 million tons) of sulfur per year.

Nitrate aerosol is not as important as sulfate aerosol, but it has the potential to become a significant source of negative forcing. One major source of nitrate aerosol is smog (the combination of ozone with oxides of nitrogen in the lower atmosphere) released from the incomplete burning of fuel in internal-combustion engines. Another source is ammonia (NH_3), which is often used in fertilizers or released by the burning of plants and other organic materials. If greater amounts of atmospheric nitrogen are converted to ammonia and agricultural ammonia emissions continue to increase as projected, the influence of nitrate aerosols on radiative forcing is expected to grow.

Both sulfate and nitrate aerosols act primarily by reflecting incoming solar radiation, thereby reducing the amount of sunlight reaching the surface. Most aerosols, unlike greenhouse gases, impart a cooling rather than warming influence on Earth's surface. One prominent exception is carbonaceous aerosols such as carbon black or soot, which are produced by the burning of fossil fuels and biomass. Carbon black tends to absorb rather than reflect incident solar radiation, and so it has a warming impact on the lower atmosphere, where it resides. Because of its absorptive properties, carbon black is also capable of having an additional indirect effect on climate. Through its deposition in snowfall, it can decrease the albedo of snow cover. This reduction in the amount of solar radiation reflected back to space by snow surfaces creates a minor positive radiative forcing.

Natural forms of aerosol include windblown mineral dust generated in arid and semiarid regions and sea salt produced by the action of waves breaking in the ocean. Changes to wind patterns as a result of climate modification could alter the emissions of these aerosols. The influence of climate change on regional patterns of aridity could shift both the sources and the destinations of dust clouds. In addition, since the concentration of sea salt aerosol, or sea aerosol, increases with the strength of the winds near the ocean surface, changes in wind speed due to global warming and climate change could influence the concentration of sea salt aerosol. For example, some studies suggest that climate change might lead to stronger winds over parts of the North Atlantic Ocean. Areas with stronger winds may experience an increase in the concentration of sea salt aerosol.

Other natural sources of aerosols include volcanic eruptions, which produce sulfate aerosol, and biogenic sources (e.g., phytoplankton), which produce dimethyl sulfide (DMS). Other important biogenic aerosols, such as terpenes, are produced naturally by certain kinds of trees or other plants. For example, the dense forests of the Blue Ridge Mountains of Virginia in the United States emit terpenes during the summer months, which in turn interact with the high humidity and warm temperatures to produce a natural photochemical smog. Anthropogenic pollutants such as nitrate and ozone, both of which serve as precursor molecules for the generation of biogenic aerosol, appear to have increased the rate of production of these aerosols severalfold. This process appears to be responsible for some of the increased aerosol pollution in regions undergoing rapid urbanization.

Human activity has greatly increased the amount of aerosol in the atmosphere compared with the background levels of preindustrial times. In contrast to the global effects of greenhouse gases, the impact of anthropogenic aerosols is confined primarily to the Northern Hemisphere, where most of the world's industrial activity occurs. The pattern of increases in anthropogenic aerosol over time is also somewhat different from that of greenhouse gases. During the middle of the 20th century, there was a substantial increase in aerosol emissions. This appears to have been at least partially responsible for a cessation of surface warming that took place in the Northern Hemisphere from the 1940s through the 1970s. Since that time, aerosol emissions have leveled off due to antipollution measures undertaken in the industrialized countries since the 1960s. Aerosol emissions may rise in the future, however, as a result of the rapid emergence of coal-fired electric power generation in China and India.

The total radiative forcing of all anthropogenic aerosols is approximately −1.2 watts per square metre. Of this total, −0.5 watt per square metre comes from direct effects (such as the reflection of solar energy back into space), and −0.7 watt per square metre comes from indirect effects (such as the influence of aerosols on cloud formation). This negative radiative forcing represents an offset of roughly 40 percent from the positive radiative forcing caused by human activity. However, the relative uncertainty in aerosol radiative forcing (approximately 90 percent) is much greater than that of greenhouse gases. In addition, future emissions of aerosols from human activities, and the influence of these emissions on future climate change, are not known with any certainty. Nevertheless, it can be said that, if concentrations of anthropogenic aerosols continue to decrease as they have since the 1970s, a significant offset to the effects of greenhouse gases will be reduced, opening future climate to further warming.

Land-use Change

There are a number of ways in which changes in land use can influence climate. The most direct influence is through the alteration of Earth's albedo, or surface reflectance. For example, the replacement of forest by cropland and pasture in the middle latitudes over the past several centuries has led to an increase in albedo, which in turn has led to greater reflection of incoming solar radiation in those regions. This replacement of forest by agriculture has been associated with a change in global average radiative forcing of approximately −0.2 watt per square metre since 1750. In Europe and other major agricultural regions, such land-use conversion began more than 1,000 years ago and has proceeded nearly to completion. For Europe, the negative radiative forcing due to land-use change has probably been substantial, perhaps approaching −5 watts per square metre. The influence of early land use on radiative forcing may help to explain a long period of cooling in Europe that followed a period of relatively mild conditions roughly 1,000 years ago. It is generally believed that the mild temperatures of this "medieval warm period," which was followed by a long period of cooling, rivaled those of 20th-century Europe.

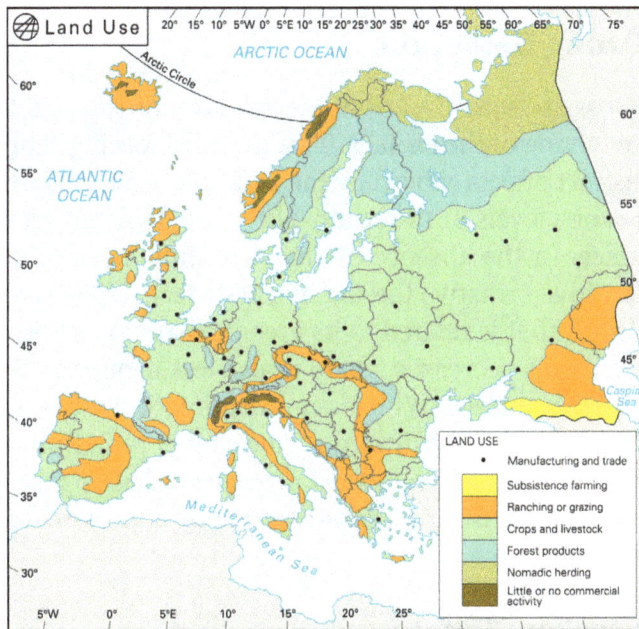

Europe: land use.

Land-use changes can also influence climate through their influence on the exchange of heat between Earth's surface and the atmosphere. For example, vegetation helps to facilitate the evaporation of water into the atmosphere through evapotranspiration. In this process, plants take up liquid water from the soil through their root systems. Eventually this water is released through transpiration into the atmosphere, as water vapour through the stomata in leaves. While deforestation generally leads to surface cooling due to the albedo factor discussed above, the land surface may also be warmed as a result of the release of latent heat by the evapotranspiration process. The relative importance of these two factors, one exerting a cooling effect and the other a warming effect, varies by both season and region. While the albedo effect is likely to dominate in middle latitudes, especially during the period from autumn through spring, the evapotranspiration effect may dominate during the summer in the midlatitudes and year-round in the tropics. The latter case is particularly important in assessing the potential impacts of continued tropical deforestation.

The rate at which tropical regions are deforested is also relevant to the process of carbon sequestration, the long-term storage of carbon in underground cavities and biomass rather than in the atmosphere. By removing carbon from the atmosphere, carbon sequestration acts to mitigate global warming. Deforestation contributes to global warming, as fewer plants are available to take up carbon dioxide from the atmosphere. In addition, as fallen trees, shrubs, and other plants are burned or allowed to slowly decompose, they release as carbon dioxide the carbon they stored during their lifetimes. Furthermore, any land-use change that influences the amount, distribution, or type of vegetation in a region can affect the concentrations of biogenic aerosols, though the impact of such changes on climate is indirect and relatively minor.

Stratospheric Ozone Depletion

Since the 1970s the loss of ozone (O_3) from the stratosphere has led to a small amount of negative radiative forcing of the surface. This negative forcing represents a competition between two distinct effects caused by the fact that ozone absorbs solar radiation. In the first case, as ozone levels in the stratosphere are depleted, more solar radiation reaches Earth's surface. In the absence of any other influence, this rise in insolation would represent a positive radiative forcing of the surface. However, there is a second effect of ozone depletion that is related to its greenhouse properties. As the amount of ozone in the stratosphere is decreased, there is also less ozone to absorb longwave radiation emitted by Earth's surface. With less absorption of radiation by ozone, there is a corresponding decrease in the downward reemission of radiation. This second effect overwhelms the first and results in a modest negative radiative forcing of Earth's surface and a modest cooling of the lower stratosphere by approximately 0.5 °C (0.9 °F) per decade since the 1970s.

Natural Influences on Climate

There are a number of natural factors that influence Earth's climate. These factors include external influences such as explosive volcanic eruptions, natural variations in the output of the Sun, and slow changes in the configuration of Earth's orbit relative to the Sun. In addition, there are natural oscillations in Earth's climate that alter global patterns of wind circulation, precipitation, and surface temperatures. One such phenomenon is the El Niño/Southern Oscillation (ENSO), a coupled atmospheric and oceanic event that occurs in the Pacific Ocean every three to seven years. In addition, the Atlantic Multidecadal Oscillation (AMO) is a similar phenomenon that occurs over decades in the North Atlantic Ocean. Other types of oscillatory behaviour that produce dramatic shifts in climate may occur across timescales of centuries and millennia.

Volcanic Aerosols

Explosive volcanic eruptions have the potential to inject substantial amounts of sulfate aerosols into the lower stratosphere. In contrast to aerosol emissions in the lower troposphere, aerosols that enter the stratosphere may remain for several years before settling out, because of the relative absence of turbulent motions there. Consequently, aerosols from explosive volcanic eruptions have the potential to affect Earth's climate. Less-explosive eruptions, or eruptions that are less vertical in orientation, have a lower potential for substantial climate impact. Furthermore, because of large-scale circulation patterns within the stratosphere, aerosols injected within tropical regions tend to spread out over the globe, whereas aerosols injected within midlatitude and polar regions tend to remain confined to the middle and high latitudes of that hemisphere. Tropical eruptions, therefore, tend to have a greater climatic impact than eruptions occurring toward the poles. In 1991 the moderate eruption of Mount Pinatubo in the Philippines provided a peak forcing of approximately −4 watts per square metre and cooled the climate by about 0.5 °C (0.9 °F) over the following few years. By comparison, the 1815 Mount Tambora eruption in present-day Indonesia, typically implicated for the 1816 "year without a summer" in Europe and North America, is believed to have been associated with a radiative forcing of approximately −6 watts per square metre.

A column of gas and ash rising from Mount Pinatubo.

While in the stratosphere, volcanic sulfate aerosol actually absorbs longwave radiation emitted by Earth's surface, and absorption in the stratosphere tends to result in a cooling of the troposphere below. This vertical pattern of temperature change in the atmosphere influences the behaviour of winds in the lower atmosphere, primarily in winter. Thus, while there is essentially a global cooling effect for the first few years following an explosive volcanic eruption, changes in the winter patterns of surface winds may actually lead to warmer winters in some areas, such as Europe. Some modern examples of major eruptions include Krakatoa (Indonesia) in 1883, El Chichón (Mexico) in 1982, and Mount Pinatubo in 1991. There is also evidence that volcanic eruptions may influence other climate phenomena such as ENSO.

Variations in Solar Output

Direct measurements of solar irradiance, or solar output, have been available from satellites only since the late 1970s. These measurements show a very small peak-to-peak variation in solar irradiance (roughly 0.1 percent of the 1,366 watts per square metre received at the top of the atmosphere, for approximately 1.4 watts per square metre).

However, indirect measures of solar activity are available from historical sunspot measurements dating back through the early 17th century. Attempts have been made to reconstruct graphs of solar irradiance variations from historical sunspot data by calibrating them against the measurements from modern satellites. However, since the modern measurements span only a few of the most recent 11-year solar cycles, estimates of solar output variability on 100-year and longer timescales are poorly correlated. Different assumptions regarding the relationship between the amplitudes of 11-year solar cycles and long-period solar output changes can lead to considerable differences in the resulting solar reconstructions. These differences in turn lead to fairly large uncertainty in estimating positive forcing by changes in solar irradiance since 1750. (Estimates range from 0.06 to 0.3 watt per square metre.) Even more challenging, given the lack of any modern analog, is the estimation of solar irradiance during the so-called Maunder Minimum, a period lasting from the mid-17th century to the early 18th century when very few sunspots were observed. While it is likely that solar irradiance was reduced at this time, it is difficult to calculate by how much. However, additional proxies of solar output exist that match reasonably well with the sunspot-derived records following the Maunder Minimum; these may be used as crude estimates of the solar irradiance variations.

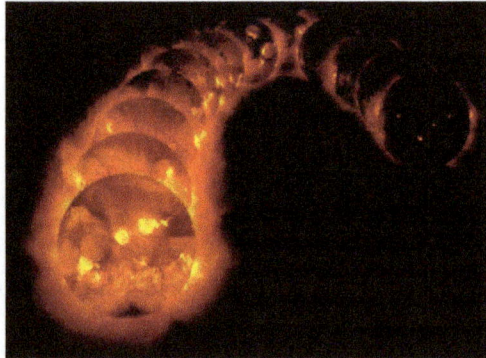

Twelve solar X-ray images obtained by Yohkoh: The solar coronal brightness decreases by a factor of about 100 during a solar cycle as the Sun goes from an "active" state (left) to a less active state (right).

In theory it is possible to estimate solar irradiance even farther back in time, over at least the past millennium, by measuring levels of cosmogenic isotopes such as carbon-14 and beryllium-10. Cosmogenic isotopes are isotopes that are formed by interactions of cosmic rays with atomic nuclei in the atmosphere and that subsequently fall to Earth, where they can be measured in the annual layers found in ice cores. Since their production rate in the upper atmosphere is modulated by changes in solar activity, cosmogenic isotopes may be used as indirect indicators of solar irradiance. However, as with the sunspot data, there is still considerable uncertainty in the amplitude of past solar variability implied by these data.

Solar forcing also affects the photochemical reactions that manufacture ozone in the stratosphere. Through this modulation of stratospheric ozone concentrations, changes in solar irradiance (particularly in the ultraviolet portion of the electromagnetic spectrum) can modify how both shortwave and longwave radiation in the lower stratosphere are absorbed. As a result, the vertical temperature profile of the atmosphere can change, and this change can in turn influence phenomena such as the strength of the winter jet streams.

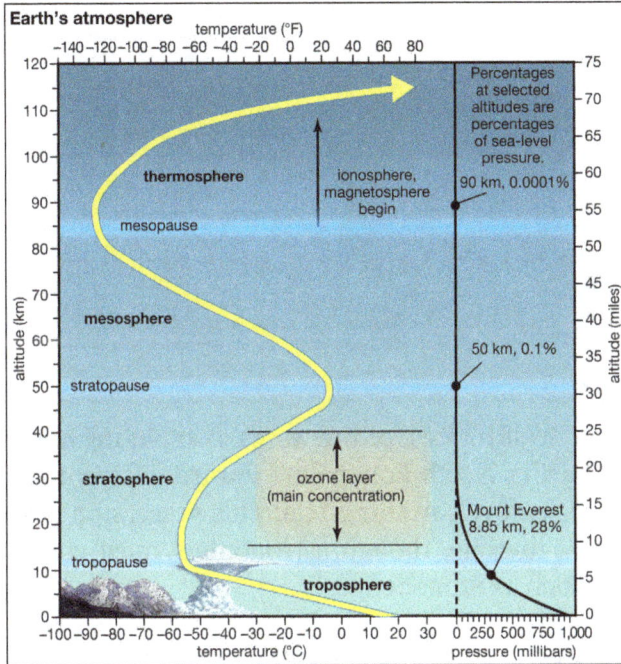

The layers of Earth's atmosphere: The yellow line shows the response of air temperature to increasing height.

Variations in Earth's Orbit

On timescales of tens of millennia, the dominant radiative forcing of Earth's climate is associated with slow variations in the geometry of Earth's orbit about the Sun. These variations include the precession of the equinoxes (that is, changes in the timing of summer and winter), occurring on a roughly 26,000-year timescale; changes in the tilt angle of Earth's rotational axis relative to the plane of Earth's orbit around the Sun, occurring on a roughly 41,000-year timescale; and changes in the eccentricity (the departure from a perfect circle) of Earth's orbit around the Sun, occurring on a roughly 100,000-year timescale. Changes in eccentricity slightly influence the mean annual solar radiation at the top of Earth's atmosphere, but the primary influence of all the orbital variations listed above is on the seasonal and latitudinal distribution of incoming solar radiation over Earth's surface. The major ice ages of the Pleistocene Epoch were closely related to the influence of these variations on summer insolation at high northern latitudes. Orbital variations thus exerted a primary control on the extent of continental ice sheets. However, Earth's orbital changes are generally believed to have had little impact on climate over the past few millennia, and so they are not considered to be significant factors in present-day climate variability.

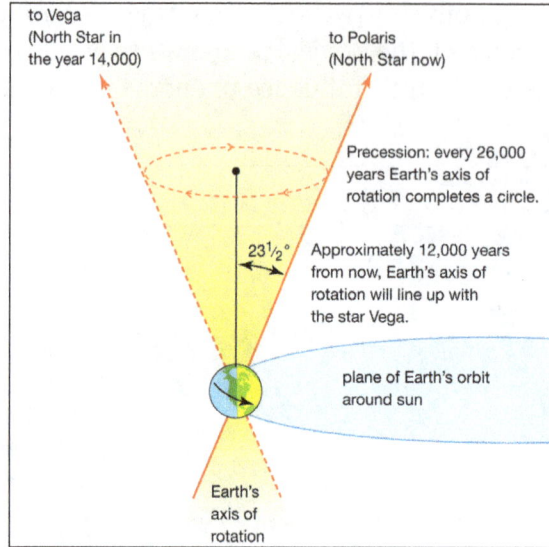

Earth's axis of rotation itself rotates, or precesses, completing one circle every 26,000 years. Consequently, Earth's North Pole points toward different stars (and sometimes toward empty space) as it travels in this circle. This precession is so slow that it is not noticeable in a person's lifetime, though astronomers must consider its effect when studying ancient sites such as Stonehenge.

Feedback Mechanisms and Climate Sensitivity

There are a number of feedback processes important to Earth's climate system and, in particular, its response to external radiative forcing. The most fundamental of these feedback mechanisms involves the loss of longwave radiation to space from the surface. Since this radiative loss increases with increasing surface temperatures according to the Stefan-Boltzmann law, it represents a stabilizing factor (that is, a negative feedback) with respect to near-surface air temperature.

Climate sensitivity can be defined as the amount of surface warming resulting from each additional watt per square metre of radiative forcing. Alternatively, it is sometimes defined as the warming that would result from a doubling of CO_2 concentrations and the associated addition of 4 watts per square metre of radiative forcing. In the absence of any additional feedbacks, climate sensitivity would be approximately 0.25 °C (0.45 °F) for each additional watt per square metre of radiative forcing. Stated alternatively, if the CO_2 concentration of the atmosphere present at the start of the industrial age (280 ppm) were doubled (to 560 ppm), the resulting additional 4 watts per square metre of radiative forcing would translate into a 1 °C (1.8 °F) increase in air temperature. However, there are additional feedbacks that exert a destabilizing, rather than stabilizing, influence, and these feedbacks tend to increase the sensitivity of climate to somewhere between 0.5 and 1.0 °C (0.9 and 1.8 °F) for each additional watt per square metre of radiative forcing.

Water Vapour Feedback

Unlike concentrations of other greenhouse gases, the concentration of water vapour in the atmosphere cannot freely vary. Instead, it is determined by the temperature of the lower atmosphere and surface through a physical relationship known as the Clausius-Clapeyron equation, named for 19th-century German physicist Rudolf Clausius and 19th-century French engineer Émile Clapeyron. Under the assumption that there is a liquid water surface in equilibrium with the atmosphere, this relationship indicates that an increase in the capacity of air to hold water vapour is a function of increasing temperature of that volume of air. This assumption is relatively good over the oceans, where water is plentiful, but not over the continents. For this reason the relative humidity (the percent of water vapour the air contains relative to its capacity) is approximately 100 percent over ocean regions and much lower over continental regions (approaching 0 percent in arid regions). Not surprisingly, the average relative humidity of Earth's lower atmosphere is similar to the fraction of Earth's surface covered by the oceans (that is, roughly 70 percent). This quantity is expected to remain approximately constant as Earth warms or cools. Slight changes to global relative humidity may result from human land-use modification, such as tropical deforestation and irrigation, which can affect the relative humidity over land areas up to regional scales.

The amount of water vapour in the atmosphere will rise as the temperature of the atmosphere rises. Since water vapour is a very potent greenhouse gas, even more potent than CO_2, the net greenhouse effect actually becomes stronger as the surface warms, which leads to even greater warming. This positive feedback is known as the "water vapour feedback." It is the primary reason that climate sensitivity is substantially greater than the previously stated theoretical value of 0.25 °C (0.45 °F) for each increase of 1 watt per square metre of radiative forcing.

Cloud Feedbacks

It is generally believed that as Earth's surface warms and the atmosphere's water vapour content increases, global cloud cover increases. However, the effects on near-surface air temperatures are complicated. In the case of low clouds, such as marine stratus clouds, the dominant radiative feature of the cloud is its albedo. Here any increase in low cloud cover acts in much the same way as an increase in surface ice cover: more incoming solar radiation is reflected and Earth's surface cools. On the other hand, high clouds, such as the towering cumulus clouds that extend up to the boundary between the troposphere and stratosphere, have a quite different impact on the surface radiation balance. The tops of cumulus clouds are considerably higher in the atmosphere and colder than their undersides. Cumulus cloud tops emit less longwave radiation out to space than the warmer cloud bottoms emit downward toward the surface. The end result of the formation of high cumulus clouds is greater warming at the surface.

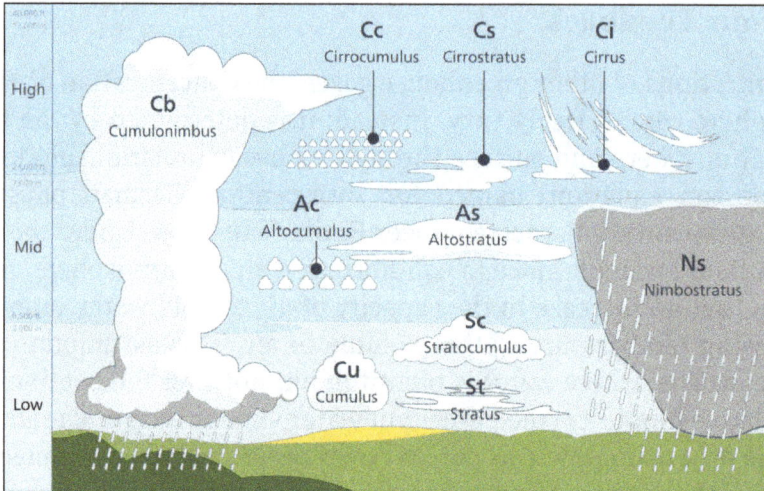

Different types of clouds form at different heights.

The net feedback of clouds on rising surface temperatures is therefore somewhat uncertain. It represents a competition between the impacts of high and low clouds, and the balance is difficult to determine. Nonetheless, most estimates indicate that clouds on the whole represent a positive feedback and thus additional warming.

Ice Albedo Feedback

Another important positive climate feedback is the so-called ice albedo feedback. This feedback arises from the simple fact that ice is more reflective (that is, has a higher albedo) than land or water surfaces. Therefore, as global ice cover decreases, the reflectivity of Earth's surface decreases, more incoming solar radiation is absorbed by the surface, and the surface warms. This feedback is considerably more important when there is relatively extensive global ice cover, such as during the height of the last ice age, roughly 25,000 years ago. On a global scale the importance of ice albedo feedback decreases as Earth's surface warms and there is relatively less ice available to be melted.

Carbon Cycle Feedbacks

Another important set of climate feedbacks involves the global carbon cycle. In particular, the two main reservoirs of carbon in the climate system are the oceans and the terrestrial biosphere. These reservoirs have historically taken up large amounts of anthropogenic CO_2 emissions. Roughly 50–70 percent is removed by the oceans, whereas the remainder is taken up by the terrestrial biosphere. Global warming, however, could decrease the capacity of these reservoirs to sequester atmospheric CO_2. Reductions in the rate of carbon uptake by these reservoirs would increase the pace of CO_2 buildup in the atmosphere and represent yet another possible positive feedback to increased greenhouse gas concentrations.

In the world's oceans, this feedback effect might take several paths. First, as surface waters warm, they would hold less dissolved CO_2. Second, if more CO_2 were added to the atmosphere and taken up by the oceans, bicarbonate ions (HCO_3^-) would multiply and ocean acidity would increase. Since calcium carbonate ($CaCO_3$) is broken down by acidic solutions, rising acidity would threaten ocean-dwelling fauna that incorporate $CaCO_3$ into their skeletons or shells. As it becomes increasingly difficult for these organisms to absorb oceanic carbon, there would be a corresponding decrease in the efficiency of the biological pump that helps to maintain the oceans as a carbon sink. Third, rising surface temperatures might lead to a slowdown in the so-called thermohaline circulation, a global pattern of oceanic flow that partly drives the sinking of surface waters near the poles and is responsible for much of the burial of carbon in the deep ocean. A slowdown in this flow due to an influx of melting fresh water into what are normally saltwater conditions might also cause the solubility pump, which transfers CO_2 from shallow to deeper waters, to become less efficient. Indeed, it is predicted that if global warming continued to a certain point, the oceans would cease to be a net sink of CO_2 and would become a net source.

As large sections of tropical forest are lost because of the warming and drying of regions such as Amazonia, the overall capacity of plants to sequester atmospheric CO_2 would be reduced. As a result, the terrestrial biosphere, though currently a carbon sink, would become a carbon source. Ambient temperature is a significant factor affecting the pace of photosynthesis in plants, and many plant species that are well adapted to their local climatic conditions have maximized their photosynthetic rates. As temperatures increase and conditions begin to exceed the optimal temperature range for both photosynthesis and soil respiration, the rate of photosynthesis would decline. As dead plants decompose, microbial metabolic activity (a CO_2 source) would increase and would eventually outpace photosynthesis.

Under sufficient global warming conditions, methane sinks in the oceans and terrestrial biosphere also might become methane sources. Annual emissions of methane by wetlands might either increase or decrease, depending on temperatures and input of nutrients, and it is possible that wetlands could switch from source to sink. There is also the potential for increased methane release as a result of the warming of Arctic permafrost (on land) and further methane release at the continental margins of the oceans (a few hundred metres below sea level). The current average atmospheric methane concentration of 1,750 ppb is equivalent to 3.5 gigatons (3.5 billion tons) of carbon. There are at least 400 gigatons of carbon equivalent stored in Arctic permafrost and as much as 10,000 gigatons (10 trillion tons) of carbon equivalent trapped on the continental margins of the oceans in a hydrated crystalline form known as clathrate. It is believed that some fraction of this trapped methane could become unstable with additional warming, although the amount and rate of potential emission remain highly uncertain.

Climate Research

Modern research into climatic variation and change is based on a variety of empirical and theoretical lines of inquiry. One line of inquiry is the analysis of data that record changes in atmosphere, oceans, and climate from roughly 1850 to the present. In a second line of inquiry, information describing paleoclimatic changes is gathered from "proxy," or indirect, sources such as ocean and lake sediments, pollen grains, corals, ice cores, and tree rings. Finally, a variety of theoretical models can be used to investigate the behaviour of Earth's climate under different conditions.

Modern Observations

Although a limited regional subset of land-based records is available from the 17th and 18th centuries, instrumental measurements of key climate variables have been collected systematically and at global scales since the mid-19th to early 20th century. These data include measurements of surface temperature on land and at sea, atmospheric pressure at sea level, precipitation over continents and oceans, sea ice extents, surface winds, humidity, and tides. Such records are the most reliable of all available climate data, since they are precisely dated and are based on well-understood instruments and physical principles. Corrections must be made for uncertainties in the data (for instance, gaps in the observational record, particularly during earlier years) and for systematic errors (such as an "urban heat island" bias in temperature measurements made on land).

Since the mid-20th century a variety of upper-air observations have become available (for example, of temperature, humidity, and winds), allowing climatic conditions to be characterized from the ground upward through the upper troposphere and lower stratosphere. Since the 1970s these data have been supplemented by polar-orbiting and geostationary satellites and by platforms in the oceans that gauge temperature, salinity, and other properties of seawater. Attempts have been made to fill the gaps in early measurements by using various statistical techniques and "backward prediction" models and by assimilating available observations into numerical weather prediction models. These techniques seek to estimate meteorological observations or atmospheric variables (such as relative humidity) that have been poorly measured in the past.

Modern measurements of greenhouse gas concentrations began with an investigation of atmospheric carbon dioxide (CO_2) concentrations by American climate scientist Charles Keeling at the summit of Mauna Loa in Hawaii in 1958. Keeling's findings indicated that CO_2 concentrations were steadily rising in association with the combustion of fossil fuels, and they also yielded the famous "Keeling curve," a graph in which the longer-term rising trend is superimposed on small oscillations related to seasonal variations in the uptake and release of CO_2 from photosynthesis and respiration in the terrestrial biosphere. Keeling's measurements at Mauna Loa apply primarily to the Northern Hemisphere.

Taking into account the uncertainties, the instrumental climate record indicates substantial trends since the end of the 19th century consistent with a warming Earth. These trends include a rise in global surface temperature of 0.9 °C (1.5 °F) between 1880 and 2012, an associated elevation of global sea level of 19–21 cm (7.5–8.3 inches) between 1901 and 2010, and a decrease in snow cover in the Northern Hemisphere of approximately 1.5 million square km (580,000 square miles). Records of average global temperatures kept by the World Meteorological Organization (WMO) indicate that the years 1998, 2005, and 2010 are statistically tied with one another as the warmest years since modern record keeping began in 1880; the WMO also noted that the decade 2001–10 was the warmest decade since 1880. Increases in global sea level are attributed to a combination of seawater expansion due to ocean heating and freshwater runoff caused by the melting of terrestrial ice. Reductions in snow cover are the result of warmer temperatures favouring a steadily shrinking winter season.

Climate data collected during the first two decades of the 21st century reveal that surface warming between 2005 and 2014 proceeded slightly more slowly than was expected from the effect of greenhouse gas increases alone. This fact was sometimes used to suggest that global warming had stopped or that it experienced a "hiatus" or "pause." In reality, this phenomenon appears to have been influenced by several factors, none of which, however, implies that global warming stopped during this period or that global warming would not continue in the future. One factor was the increased burial of heat beneath the ocean surface by strong trade winds, a process assisted by La Niña conditions. The effects of La Niña manifest in the form of cooling surface waters along the western coast of South America. As a result, warming at the ocean surface was reduced, but the accumulation of heat in other parts of the ocean occurred at an accelerated rate. Another factor cited by climatologists was a small but potentially important increase in aerosols from volcanic activity, which may have blocked a small portion of incoming solar radiation and which were accompanied by a small reduction in solar output during the period. These factors, along with natural decades-long oscillations in the climate system, may have masked a portion of the greenhouse warming. (However, climatologists point out that these natural climate cycles are expected to add to greenhouse warming in the future when the oscillations eventually reverse direction.) For these reasons many scientists believe that it is an error to call this slowdown in detectable surface warming a "hiatus" or a "pause."

Climate Records

In order to reconstruct climate changes that occurred prior to about the mid-19th century, it is necessary to use "proxy" measurements—that is, records of other natural phenomena that indirectly measure various climate conditions. Some proxies, such as most sediment cores and pollen records, glacial moraine evidence, and geothermal borehole temperature profiles, are coarsely resolved or dated and thus are only useful for describing climate changes on long timescales. Other proxies, such as growth rings from trees or oxygen isotopes from corals and ice cores, can provide a record of yearly or even seasonal climate changes.

The data from these proxies should be calibrated to known physical principles or related statistically to the records collected by modern instruments, such as satellites.

Networks of proxy data can then be used to infer patterns of change in climate variables, such as the behaviour of surface temperature over time and geography. Yearly reconstructions of climate variables are possible over the past 1,000 to 2,000 years using annually dated proxy records, but reconstructions farther back in time are generally based on more coarsely resolved evidence such as ocean sediments and pollen records. For these, records of conditions can be reconstructed only on timescales of hundreds or thousands of years. In addition, since relatively few long-term proxy records are available for the Southern Hemisphere, most reconstructions focus on the Northern Hemisphere.

The various proxy-based reconstructions of the average surface temperature of the Northern Hemisphere differ in their details. These differences are the result of uncertainties implicit in the proxy data themselves and also of differences in the statistical methods used to relate the proxy data to surface temperature. Nevertheless, all studies as reviewed in the IPCC's Fourth Assessment Report (AR4), which was published in 2007, indicate that the average surface temperature since about 1950 is higher than at any time during the previous 1,000 years.

Theoretical Climate Models

Theoretical models of Earth's climate system can be used to investigate the response of climate to external radiative forcing as well as its own internal variability. Two or more models that focus on different physical processes may be coupled or linked together through a common feature, such as geographic location. Climate models vary considerably in their degree of complexity. The simplest models of energy balance describe Earth's surface as a globally uniform layer whose temperature is determined by a balance of incoming and outgoing shortwave and longwave radiation. These simple models may also consider the effects of greenhouse gases. At the other end of the spectrum are fully coupled, three-dimensional, global climate models. These are complex models that solve for radiative balance; for laws of motion governing the atmosphere, ocean, and ice; and for exchanges of energy and momentum within and between the different components of the climate. In some cases, theoretical climate models also include an interactive representation of Earth's biosphere and carbon cycle.

Even the most-detailed climate models cannot resolve all the processes that are important in the atmosphere and ocean. Most climate models are designed to gauge the behaviour of a number of physical variables over space and time, and they often artificially divide Earth's surface into a grid of many equal-sized "cells." Each cell may neatly correspond to some physical process (such as summer near-surface air temperature) or other variable (such as land-use type), and it may be assigned a relatively straightforward value. So-called "sub-grid-scale" processes, such as those of clouds, are too small to be captured by the relatively coarse spacing of the individual grid cells. Instead, such processes must be represented through a statistical process that relates the properties of the atmosphere and ocean. For example, the average fraction of cloud cover over a hypothetical "grid box" (that is, a representative volume of air or water in the model) can be estimated from the average relative humidity and the vertical temperature profile of the grid cell. Variations in the behaviour of different coupled climate models arise in large part from differences in the ways sub-grid-scale processes are mathematically expressed.

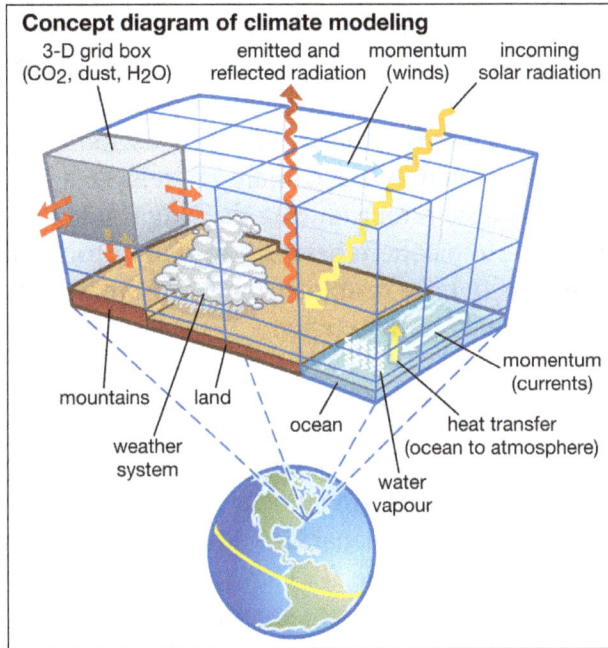

Concept diagram of climate modeling

To understand and explain the complex behaviour of Earth's climate, modern climate models incorporate several variables that stand in for materials passing through Earth's atmosphere and oceans and the forces that affect them.

Despite these required simplifications, many theoretical climate models perform remarkably well when reproducing basic features of the atmosphere, such as the behaviour of midlatitude jet streams or Hadley cell circulation. The models also adequately reproduce important features of the oceans, such as the Gulf Stream. In addition, models are becoming better able to reproduce the main patterns of internal climate variability, such as those of El Niño/Southern Oscillation (ENSO). Consequently, periodically recurring events—such as ENSO and other interactions between the atmosphere and ocean currents—are being modeled with growing confidence.

Climate models have been tested in their ability to reproduce observed changes in response to radiative forcing. In 1988 a team at NASA's Goddard Institute for Space Studies in New York City used a fairly primitive climate model to predict warming patterns that might occur in response to three different scenarios of anthropogenic radiative forcing. Warming patterns were forecast for subsequent decades. Of the three scenarios, the middle one, which corresponds most closely to actual historical carbon emissions, comes closest to matching the observed warming of roughly

0.5 °C (0.9 °F) that has taken place since then. The NASA team also used a climate model to successfully predict that global mean surface temperatures would cool by about 0.5 °C for one to two years after the 1991 eruption of Mount Pinatubo in the Philippines.

More recently, so-called "detection and attribution" studies have been performed. These studies compare predicted changes in near-surface air temperature and other climate variables with patterns of change that have been observed for the past one to two centuries. The simulations have shown that the observed patterns of warming of Earth's surface and upper oceans, as well as changes in other climate phenomena such as prevailing winds and precipitation patterns, are consistent with the effects of an anthropogenic influence predicted by the climate models. In addition, climate model simulations have shown success in reproducing the magnitude and the spatial pattern of cooling in the Northern Hemisphere between roughly 1400 and 1850—during the Little Ice Age, which appears to have resulted from a combination of lowered solar output and heightened explosive volcanic activity.

Potential Effects of Global Warming

The path of future climate change will depend on what courses of action are taken by society—in particular the emission of greenhouse gases from the burning of fossil fuels. A range of alternative emissions scenarios known as representative concentration pathways (RCPs) were proposed by the IPCC in the Fifth Assessment Report (AR5), which was published in 2014, to examine potential future climate changes. The scenarios depend on various assumptions concerning future rates of human population growth, economic development, energy demand, technological advancement, and other factors. Unlike the scenarios used in previous IPCC assessments, the AR5 RCPs explicitly account for climate change mitigation efforts.

Projected range of sea-level rise by climate change scenario		
Scenario	Temperature change (°C) in 2090–99 relative to 1980–99	Sea-level rise (m) in 2090–99 relative to 1980–99
B1	1.1–2.9	0.18–0.38
A1T	1.4–3.8	0.20–0.45
B2	1.4–3.8	0.20–0.43
A1B	1.7–4.4	0.21–0.48
A2	2.0–5.4	0.23–0.51
A1Fl	2.4–6.4	0.26–0.59
Ranges of sea-level rise are based on various models of climate change that exclude the possibility of future rapid changes in ice flow, such as the melting of the Greenland and Antarctic ice caps.		

The results of each scenario in the IPCC's Fourth Assessment Report are depicted in the graph below.

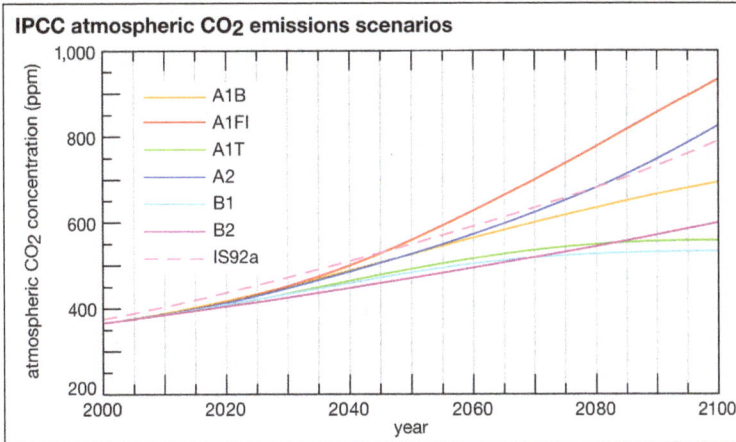

Graph of the predicted increase in the concentration of carbon dioxide (CO_2) in Earth's atmosphere according to a series of climate change scenarios that assume different levels of economic development, population growth, and fossil fuel use.

The AR5 scenario with the smallest increases in greenhouse gases is RCP 2.6, which denotes the net radiative forcing by 2100 in watts per square metre (a doubling of CO_2 concentrations from preindustrial values of 280 ppm to 560 ppm represents roughly 3.7 watts per square metre). RCP 2.6 assumes substantial improvements in energy efficiency, a rapid transition away from fossil fuel energy, and a global population that peaks at roughly nine billion people in the 21st century. In that scenario CO_2 concentrations remain below 450 ppm and actually fall toward the end of the century (to about 420 ppm) as a result of widespread deployment of carbon-capture technology.

Scenario RCP 8.5, by contrast, might be described as "business as usual." It reflects the assumption of an energy-intensive global economy, high population growth, and a reduced rate of technological development. CO_2 concentrations are more than three times greater than preindustrial levels (roughly 936 ppm) by 2100 and continue to grow thereafter. RCP 4.5 and RCP 6.0 envision intermediate policy choices, resulting in stabilization by 2100 of CO_2 concentrations at 538 and 670 ppm, respectively. In all those scenarios, the cooling effect of industrial pollutants such as sulfate particulates, which have masked some of the past century's warming, is assumed to decline to near zero by 2100 because of policies restricting their industrial production.

Simulations of Future Climate Change

The differences between the various simulations arise from disparities between the various climate models used and from assumptions made by each emission scenario. For example, best estimates of the predicted increases in global surface temperature between the years 2000 and 2100 range from about 0.3 to 4.8 °C (0.5 to 8.6 °F), depending on which emission scenario is assumed and which climate model is used. Relative to preindustrial (i.e., 1750–1800) temperatures, these estimates reflect an overall warming of the globe of

1.4 to 5.0 °C (2.5 to 9.0 °F). These projections are conservative in that they do not take into account potential positive carbon cycle feedbacks. Only the lower-end emissions scenario RCP 2.6 has a reasonable chance (roughly 50 percent) of holding additional global surface warming by 2100 to less than 2.0 °C (3.6 °F)—a level considered by many scientists to be the threshold above which pervasive and extreme climatic effects will occur.

Patterns of Warming

The greatest increase in near-surface air temperature is projected to occur over the polar region of the Northern Hemisphere because of the melting of sea ice and the associated reduction in surface albedo. Greater warming is predicted over land areas than over the ocean. Largely due to the delayed warming of the oceans and their greater specific heat, the Northern Hemisphere—with less than 40 percent of its surface area covered by water—is expected to warm faster than the Southern Hemisphere. Some of the regional variation in predicted warming is expected to arise from changes to wind patterns and ocean currents in response to surface warming. For example, the warming of the region of the North Atlantic Ocean just south of Greenland is expected to be slight. This anomaly is projected to arise from a weakening of warm northward ocean currents combined with a shift in the jet stream that will bring colder polar air masses to the region.

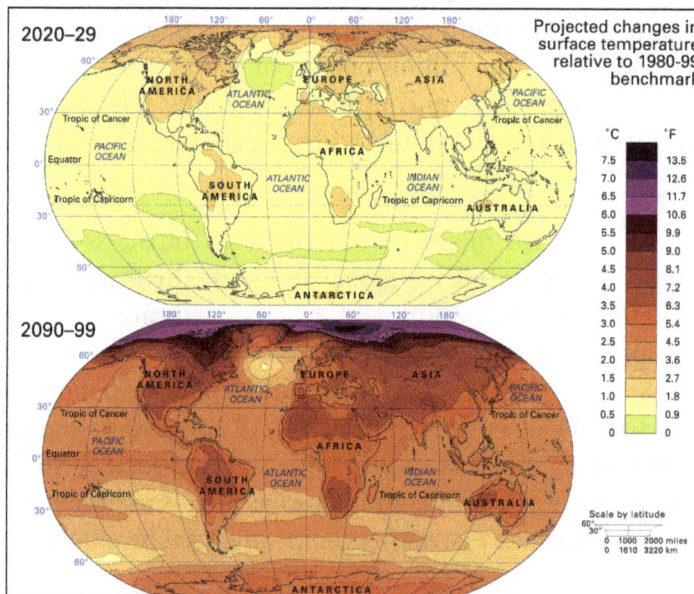

Projected changes in mean surface temperatures by the late 21st century according to the A1B climate change scenario.

Precipitation Patterns

The climate changes associated with global warming are also projected to lead to changes in precipitation patterns across the globe. Increased precipitation is predicted in the polar and subpolar regions, whereas decreased precipitation is projected for the middle latitudes of both hemispheres as a result of the expected poleward shift in the jet streams. Whereas precipitation near the Equator is predicted to increase, it is thought that rainfall in the subtropics will decrease. Both phenomena are associated with a forecasted strengthening of the tropical Hadley cell pattern of atmospheric circulation.

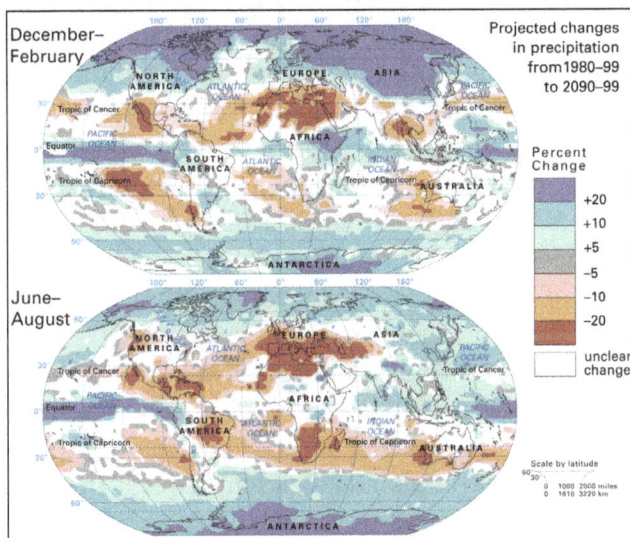

Projected changes in mean precipitation by the late 21st century according to the A1B climate change scenario.

Changes in precipitation patterns are expected to increase the chances of both drought and flood conditions in many areas. Decreased summer precipitation in North America, Europe, and Africa, combined with greater rates of evaporation due to warming surface temperatures, is projected to lead to decreased soil moisture and drought in many regions. Furthermore, since anthropogenic climate change will likely lead to a more vigorous hydrologic cycle with greater rates of both evaporation and precipitation, there will be a greater probability for intense precipitation and flooding in many regions.

Regional Predictions

Regional predictions of future climate change remain limited by uncertainties in how the precise patterns of atmospheric winds and ocean currents will vary with increased surface warming. For example, some uncertainty remains in how the frequency and magnitude of El Niño/Southern Oscillation (ENSO) events will adjust to climate change. Since ENSO is one of the most prominent sources of interannual variations in regional patterns of precipitation and temperature, any uncertainty in how it will change implies a corresponding uncertainty in certain regional patterns of climate change. For example, increased El Niño activity would likely lead to more winter precipitation in some regions, such as the desert southwest of the United States. This might offset the drought predicted for those regions, but at the same time it might lead to less precipitation in other regions. Rising winter precipitation in the desert southwest of the United States might exacerbate drought conditions in locations as far away as South Africa.

Ice Melt and Sea Level Rise

A warming climate holds important implications for other aspects of the global environment. Because of the slow process of heat diffusion in water, the world's oceans are likely to continue to warm for several centuries in response to increases in greenhouse concentrations that have taken place so far. The combination of seawater's thermal expansion associated with this warming and the melting of mountain glaciers is predicted to lead to an increase in global sea level of 0.45–0.82 metre (1.4–2.7 feet) by 2100 under the RCP 8.5 emissions scenario. However, the actual rise in sea level could be considerably greater than this. It is probable that the continued warming of Greenland will cause its ice sheet to melt at accelerated rates. In addition, this level of surface warming may also melt the ice sheet of West Antarctica. Paleoclimatic evidence suggests that an additional 2 °C (3.6 °F) of warming could lead to the ultimate destruction of the Greenland Ice Sheet, an event that would add another 5 to 6 metres (16 to 20 feet) to predicted sea level rise. Such an increase would submerge a substantial number of islands and lowland regions. Coastal lowland regions vulnerable to sea level rise include substantial parts of the U.S. Gulf Coast and Eastern Seaboard (including roughly the lower third of Florida), much of the Netherlands and Belgium (two of the European Low Countries), and heavily populated tropical areas such as Bangladesh. In addition, many of the world's major cities—such as Tokyo, New York, Mumbai, Shanghai, and Dhaka—are located in lowland regions vulnerable to rising sea levels. With the loss of the West Antarctic ice sheet, additional sea level rise would approach 10.5 metres (34 feet).

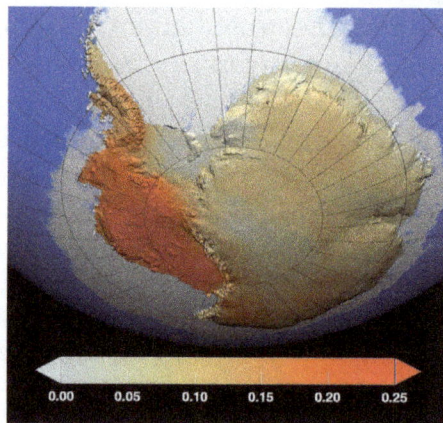

NASA image showing locations on Antarctica where temperatures: Red represents areas where temperatures had increased the most over the period, particularly in West Antarctica, while dark blue represents areas with a lesser degree of warming. Temperature changes are measured in degrees Celsius.

While the current generation of models predicts that such global sea level changes might take several centuries to occur, it is possible that the rate could accelerate as a result of processes that tend to hasten the collapse of ice sheets. One such process is the development of moulins—large vertical shafts in the ice that allow surface meltwater to penetrate to the base of the ice sheet. A second process involves the vast ice shelves off Antarctica that buttress the grounded continental ice sheet of Antarctica's interior. If those ice shelves collapse, the continental ice sheet could become unstable, slide rapidly toward the ocean, and melt, thereby further increasing mean sea level. Thus far, neither process has been incorporated into the theoretical models used to predict sea level rise.

Ocean Circulation Changes

Another possible consequence of global warming is a decrease in the global ocean circulation system known as the "thermohaline circulation" or "great ocean conveyor belt." This system involves the sinking of cold saline waters in the subpolar regions of the oceans, an action that helps to drive warmer surface waters poleward from the subtropics. As a result of this process, a warming influence is carried to Iceland and the coastal regions of Europe that moderates the climate in those regions. Some scientists believe that global warming could shut down this ocean current system by creating an influx of fresh water from melting ice sheets and glaciers into the subpolar North Atlantic Ocean. Since fresh water is less dense than saline water, a significant intrusion of fresh water would lower the density of the surface waters and thus inhibit the sinking motion that drives the large-scale thermohaline circulation. It has also been speculated that, as a consequence of large-scale surface warming, such changes could even trigger colder conditions in regions surrounding the North Atlantic. Experiments with modern climate models suggest that such an event would be unlikely. Instead, a moderate weakening of the thermohaline circulation might occur that would lead to a dampening of surface warming—rather than actual cooling—in the higher latitudes of the North Atlantic Ocean.

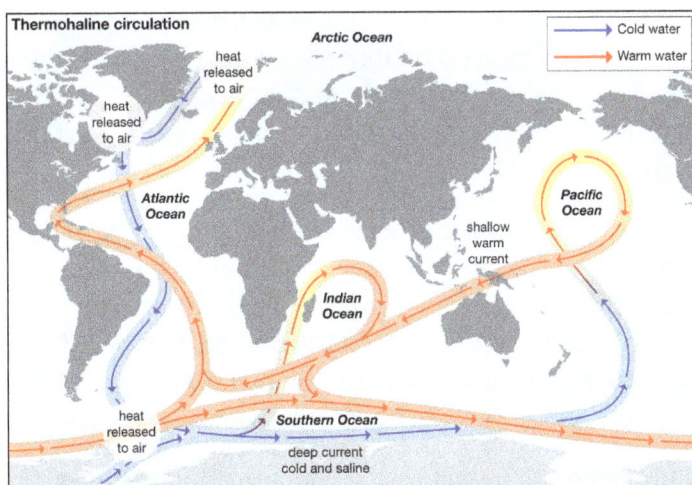

Thermohaline circulation transports and mixes the water of the oceans. In the process it transports heat, which influences regional climate patterns. The density of seawater is determined by the temperature and salinity of a volume of seawater at a particular location. The difference in density between one location and another drives the thermohaline circulation.

Tropical Cyclones

One of the more controversial topics in the science of climate change involves the impact of global warming on tropical cyclone activity. It appears likely that rising tropical ocean temperatures associated with global warming will lead to an increase in the intensity (and the associated destructive potential) of tropical cyclones. In the Atlantic a close relationship has been observed between rising ocean temperatures and a rise in the strength of hurricanes. Trends in the intensities of tropical cyclones in other regions, such as in the tropical Pacific and Indian oceans, are more uncertain due to a paucity of reliable long-term measurements.

While the warming of oceans favours increased tropical cyclone intensities, it is unclear to what extent rising temperatures affect the number of tropical cyclones that occur each year. Other factors, such as wind shear, could play a role. If climate change increases the amount of wind shear—a factor that discourages the formation of tropical cyclones—in regions where such storms tend to form, it might partially mitigate the impact of warmer temperatures. On the other hand, changes in atmospheric winds are themselves uncertain—because of, for example, uncertainties in how climate change will affect ENSO.

Environmental Consequences of Global Warming

Global warming and climate change have the potential to alter biological systems. More specifically, changes to near-surface air temperatures will likely influence ecosystem functioning and thus the biodiversity of plants, animals, and other forms of life. The current geographic ranges of plant and animal species have been established by adaptation to long-term seasonal climate patterns. As global warming alters these patterns on timescales considerably shorter than those that arose in the past from natural climate variability, relatively sudden climatic changes may challenge the natural adaptive capacity of many species.

A large fraction of plant and animal species are likely to be at an increased risk of extinction if global average surface temperatures rise another 1.5 to 2.5 °C (2.7 to 4.5 °F) by the year 2100. Species loss estimates climb to as much as 40 percent for a warming in excess of 4.5 °C (8.1 °F)—a level that could be reached in the IPCC's higher emissions scenarios. A 40 percent extinction rate would likely lead to major changes in the food webs within ecosystems and have a destructive impact on ecosystem function.

Surface warming in temperate regions is likely to lead changes in various seasonal processes—for instance, earlier leaf production by trees, earlier greening of vegetation, altered timing of egg laying and hatching, and shifts in the seasonal migration patterns of birds, fishes, and other migratory animals. In high-latitude ecosystems, changes in the seasonal patterns of sea ice threaten predators such as polar bears and walruses; both species rely on broken sea ice for their hunting activities. Also in the high latitudes, a combination of warming waters, decreased sea ice, and changes in ocean salinity and circulation is likely to lead to reductions or redistributions in populations of algae and plankton. As a result, fish and other organisms that forage upon algae and plankton may be threatened. On land, rising temperatures and changes in precipitation patterns and drought frequencies are likely to alter patterns of disturbance by fires and pests.

Numerous ecologists, conservation biologists, and other scientists studying climate warn that rising surface temperatures will bring about an increased extinction risk. In 2015 one study that examined 130 extinction models developed in previous studies predicted that 5.2 percent of species would be lost with a rise in average temperatures of 2 °C (3.6 °F) above temperature benchmarks from before the onset of the Industrial Revolution. The study also predicted that 16 percent of Earth's species would be lost if surface warming increased to about 4.3 °C (7.7 °F) above preindustrial temperature benchmarks.

Other likely impacts on the environment include the destruction of many coastal wetlands, salt marshes, and mangrove swamps as a result of rising sea levels and the loss of certain rare and fragile habitats that are often home to specialist species that are unable to thrive in other environments. For example, certain amphibians limited to isolated tropical cloud forests either have become extinct already or are under serious threat of extinction. Cloud forests—tropical forests that depend on persistent condensation of moisture in the air—are disappearing as optimal condensation levels move to higher elevations in response to warming temperatures in the lower atmosphere.

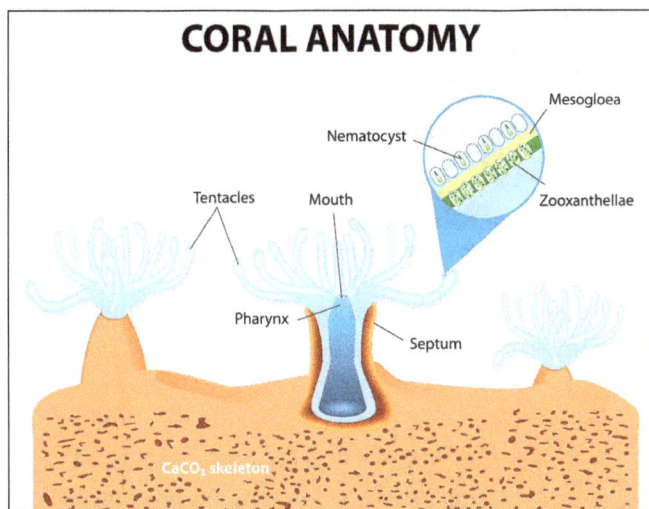

Cross section of a generalized coral polyp.

In many cases a combination of stresses caused by climate change as well as human activity represents a considerably greater threat than either climatic stresses or nonclimatic stresses alone. A particularly important example is coral reefs, which contain much of the ocean's biodiversity. Rising ocean temperatures increase the tendency for coral bleaching (a condition where zooxanthellae, or yellow-green algae, living in symbiosis with coral either lose their pigments or abandon the coral polyps altogether), and they also raise the likelihood of greater physical damage by progressively more destructive tropical cyclones. In many areas coral is also under stress from increased ocean acidification, marine pollution, runoff from agricultural fertilizer, and physical damage by boat anchors and dredging.

Another example of how climate and nonclimatic stresses combine is illustrated by the threat to migratory animals. As these animals attempt to relocate to regions with more favourable climate conditions, they are likely to encounter impediments such as highways, walls, artificial waterways, and other man-made structures.

Warmer temperatures are also likely to affect the spread of infectious diseases, since the geographic ranges of carriers, such as insects and rodents, are often limited by climatic conditions. Warmer winter conditions in New York in 1999, for example, appear to have facilitated an outbreak of West Nile virus, whereas the lack of killing frosts in New Orleans during the early 1990s led to an explosion of disease-carrying mosquitoes and cockroaches. Warmer winters in the Korean peninsula and southern Europe have allowed the spread of the Anopheles mosquito, which carries the malaria parasite, whereas warmer conditions in Scandinavia in recent years have allowed for the northward advance of encephalitis.

Anopheles mosquito, carrier of the malarial parasite.

In the southwestern United States, alternations between drought and flooding related in part to the ENSO phenomenon have created conditions favourable for the spread of hantaviruses by rodents. The spread of mosquito-borne Rift Valley fever in equatorial East Africa has also been related to wet conditions in the region associated with ENSO. Severe weather conditions conducive to rodents or insects have been implicated in infectious disease outbreaks—for instance, the outbreaks of cholera and leptospirosis that occurred after Hurricane Mitch struck Central America in 1998. Global warming could therefore affect the spread of infectious disease through its influence on ENSO or on severe weather conditions.

Socioeconomic Consequences of Global Warming

Socioeconomic impacts of global warming could be substantial, depending on the actual temperature increases over the next century. Models predict that a net global warming of 1 to 3 °C (1.8 to 5.4 °F) beyond the late 20th-century global average would produce economic losses in some regions (particularly the tropics and high latitudes) and economic benefits in others. For warming beyond those levels, benefits would tend to decline and costs increase. For warming in excess of 4 °C (7.2 °F), models predict that costs will exceed benefits on average, with global mean economic losses estimated between 1 and 5 percent of gross domestic product. Substantial disruptions could be expected under those conditions, specifically in the areas of agriculture, food and forest products, water and energy supply, and human health.

Agricultural productivity might increase modestly in temperate regions for some crops in response to a local warming of 1–3 °C (1.8–5.4 °F), but productivity will generally decrease with further warming. For tropical and subtropical regions, models predict decreases in crop productivity for even small increases in local warming. In some cases, adaptations such as altered planting practices are projected to ameliorate losses in productivity for modest amounts of warming. An increased incidence of drought and flood events would likely lead to further decreases in agricultural productivity and to decreases in livestock production, particularly among subsistence farmers in tropical regions. In regions such as the African Sahel, decreases in agricultural productivity have already been observed as a result of shortened growing seasons, which in turn have occurred as a result of warmer and drier climatic conditions. In other regions, changes in agricultural practice, such as planting crops earlier in the growing season, have been undertaken. The warming of oceans is predicted to have an adverse impact on commercial fisheries by changing the distribution and productivity of various fish species, whereas commercial timber productivity may increase globally with modest warming.

Water resources are likely to be affected substantially by global warming. At current rates of warming, a 10–40 percent increase in average surface runoff and water availability has been projected in higher latitudes and in certain wet regions in the tropics by the middle of the 21st century, while decreases of similar magnitude are expected in other parts of the tropics and in the dry regions in the subtropics. This would be particularly severe during the summer season. In many cases water availability is already decreasing or expected to decrease in regions that have been stressed for water resources since the turn of the 21st century. Such regions as the African Sahel, western North America, southern Africa, the Middle East, and western Australia continue to be particularly vulnerable. In these regions drought is projected to increase in both magnitude and extent, which would bring about adverse effects on agriculture and livestock raising. Earlier and increased spring runoff is already being observed in western North America and other temperate regions served by glacial or snow-fed streams and rivers. Fresh water currently stored by mountain glaciers and snow in both the tropics and extratropics is also projected to decline and thus reduce the availability of fresh water for more than 15 percent of the world's population. It is also likely that warming temperatures, through their impact on biological activity in lakes and rivers, may have an adverse impact on water quality, further diminishing access to safe water sources for drinking or farming. For example, warmer waters favour an increased frequency of nuisance algal blooms, which can pose health risks to humans. Risk-management procedures have already been taken by some countries in response to expected changes in water availability.

Energy availability and use could be affected in at least two distinct ways by rising surface temperatures. In general, warmer conditions would favour an increased demand for air-conditioning; however, this would be at least partially offset by decreased demand for winter heating in temperate regions. Energy generation that requires water either directly, as in hydroelectric power, or indirectly, as in steam turbines used in coal-fired power plants or in cooling towers used in nuclear power plants, may become more difficult in regions with reduced water supplies.

As discussed above, it is expected that human health will be further stressed under global warming conditions by potential increases in the spread of infectious diseases. Declines in overall human health might occur with increases in the levels of malnutrition due to disruptions in food production and by increases in the incidence of afflictions. Such afflictions could include diarrhea, cardiorespiratory illness, and allergic reactions in the midlatitudes of the Northern Hemisphere as a result of rising levels of pollen. Rising heat-related mortality, such as that observed in response to the 2003 European heat wave, might occur in many regions, especially in impoverished areas where air-conditioning is not generally available.

The economic infrastructure of most countries is predicted to be severely strained by global warming and climate change. Poor countries and communities with limited adaptive capacities are likely to be disproportionately affected. Projected increases in the incidence of severe weather, heavy flooding, and wildfires associated with reduced summer ground moisture in many regions will threaten homes, dams, transportation networks and other facets of human infrastructure. In high-latitude and mountain regions, melting permafrost is likely to lead to ground instability or rock avalanches, further threatening structures in those regions. Rising sea levels and the increased potential for severe tropical cyclones represent a heightened threat to coastal communities throughout the world. It has been estimated that an additional warming of 1–3 °C (1.8–5.4 °F) beyond the late 20th-century global average would threaten millions more people with the risk of annual flooding. People in the densely populated, poor, low-lying regions of Africa, Asia, and tropical islands would be the most vulnerable, given their limited adaptive capacity. In addition, certain regions in developed countries, such as the Low Countries of Europe and the Eastern Seaboard and Gulf Coast of the United States, would also be vulnerable to the effects of rising sea levels. Adaptive steps are already being taken by some governments to reduce the threat of increased coastal vulnerability through the construction of dams and drainage works.

Climate Change

The climate can be described as the average weather over a period of time. Climate change means a significant change in the measures of climate, such as temperature, rainfall, or wind, lasting for an extended period – decades or longer. The Earth's climate has changed many times during the planet's history, with events ranging from ice ages to long periods of warmth. What's different about this period of the earth's history is that human activities are significantly contributing to natural climate change through our emissions of greenhouse gases. This interference is resulting in increased air and ocean temperatures, drought, melting ice and snow, rising sea levels, increased rainfall, flooding and other influences.

Causes of Climate Change

Climate change can result from natural processes and factors and more recently due to human activities through our emissions of greenhouse gases. Examples of natural factors include:

- Changes in the sun's intensity.

- Volcanic eruptions, or slow changes in the Earth's orbit around the sun.

- Natural processes within the climate system such as changes in ocean current circulation.

However, the current global aim is to tackle climate change resulting from human activities whose greenhouse gas emissions are changing the composition of the earth's atmosphere. The Intergovernmental Panel on Climate Change (IPCC) state that:

"Most of the observed increase in global average temperatures since the mid-20th century is very likely due to the observed increase in anthropogenic (produced by humans) greenhouse gas emissions".

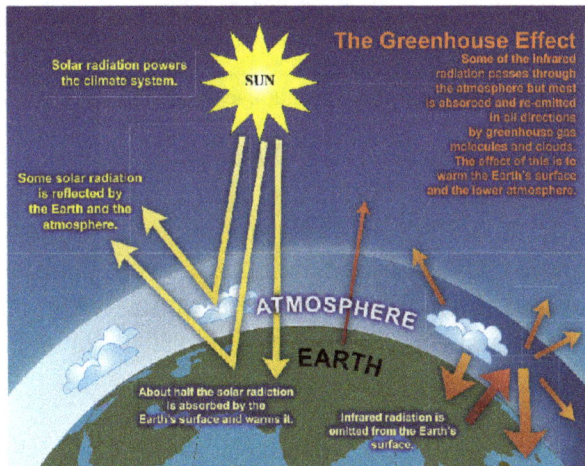

The greenhouse gas effect.

Examples of human activities contributing to climate change include:

- Carbon dioxide emissions through burning fossil fuels such as coal, oil and gas and peat.

- Methane and nitrous oxide emissions from agriculture.

- Emissions through land use changes such as deforestation, reforestation, urbanization, and desertification.

These emissions that are changing the composition of the earths atmosphere are termed the Greenhouse effect. For the past 200 years, the burning of fossil fuels, such as coal and oil, and deforestation have caused the concentrations of heat-trapping greenhouse gases to increase significantly in our atmosphere. These gases prevent heat from escaping to space, somewhat like the glass panels of a greenhouse. The figure above from the IPCC explains greenhouse gases.

The increased emissions to the atmosphere of greenhouse gases mean that current levels of gases far exceed their natural ranges. The figure below shows the levels of certain greenhouse gases in the atmosphere over the last 2000 years.

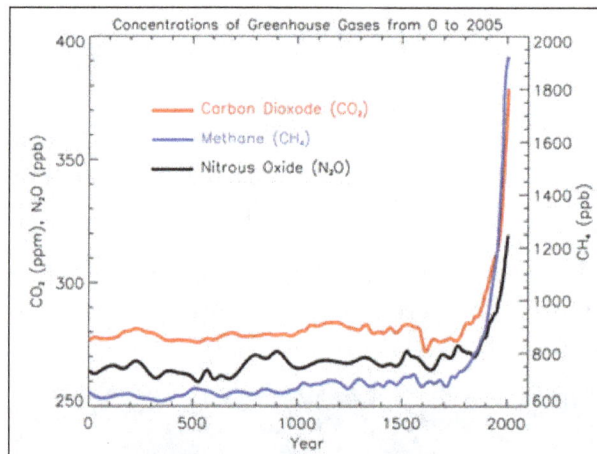

Atmospheric Greenhouse Gas Level.

This rise in greenhouse gases has increased the amount of energy being trapped in the climate system. The consequences of this are most clearly evident in the global temperature records, which show that, on average, the global temperature has increased by 0.8 degrees centigrade (°C) above pre-industrial levels. Continued emissions at or above current levels would cause further warming and result in changes in the global climate system during the 21st century that would very likely be larger than those observed during the 20th century.

Chapter 2

Climate Change

Climate change refers to the phenomena where changes in the Earth's climate system lead to new weather patterns that stay for a substantial period of time. It can be caused due to natural reasons as well as human activities. This chapter closely examines these key causes of climate change as well as the external force mechanisms to provide an extensive understanding of the subject.

Effects of Climate Change

Climate change destabilises the Earth's temperature equilibrium and has far-reaching effects on human beings and the environment. During the course of global warming, the energy balance and thus the temperature of the earth change, due to the increased concentration of greenhouse gases, which has a significant impact on humans and the environment.

It is not scientifically possible to assign individual weather events to the current climate change, however, it can be statistically proven that global warming will increase the probability of extreme weather events.

The direct consequences of man-made climate change include:

- Rising maximum temperatures.
- Rising minimum temperatures.
- Rising sea levels.
- Higher ocean temperatures.
- An increase in heavy precipitation (heavy rain and hail).
- Shrinking glaciers.
- Thawing permafrost.

The indirect consequences of climate change, which directly affect us humans and our environment, include:

- An increase in hunger and water crises, especially in developing countries.
- Health risks through rising air temperatures and heatwaves.

- Economic implications of dealing with secondary damage related to climate change.

- Increasing spread of pests and pathogens.

- Loss of biodiversity due to limited adaptability and adaptability speed of flora and fauna.

- Ocean acidification due to increased hco3 concentrations in the water as a consequence of increased CO_2 concentrations.

- The need for adaptation in all areas (e.g. Agriculture, forestry, energy, infrastructure, tourism, etc.).

As the global climate is a highly interconnected system that is influenced by many different factors, the consequences usually result in positive or negative feedback effects. This refers to developments that are self-enhancing due to the occurrence of certain conditions.

A common example is the ice-albedo feedback, which refers to the melting of the polar caps. According to this, extensive ice surfaces have a cooling effect on the global climate, as a high proportion of radiation is reflected. As a result of the global rise in the average temperature, however, these ice surfaces begin to melt, the ice surfaces shrink and the amount of reflected radiation is reduced. At the same time, the area of land or ocean that has a significantly lower albedo will increase, reflecting less radiation and thus intensifying the actual cause of glacier melt.

Furthermore, scientists can calculate the so-called tipping points of individual subsystems of the global climate. The higher the global rise in temperature, the more the climate system is affected, so that at a certain point, despite significant efforts, a reversal in the process is no longer possible. Where exactly these tipping points can be found, however, is currently still unclear and can only be calculated with a great degree of uncertainty. Such tipping points are expected for the melting of the polar caps and for the stability of important ocean currents.

Natural Causes of Climate Change

The natural variability and the climate fluctuations of the climate system have always been part of the Earth's history. To understand climate change fully, the causes of climate change must be first identified. The earth's climate is influenced and changed through natural causes like volcanic eruptions, ocean currents, the Earth's orbital changes, solar variations and internal variability.

Volcanic eruptions - The main effect volcanoes have on the climate is short-term cooling. Volcanic eruptions pump out clouds of dust and ash, which block out some sunlight.

Because the ash particles are relatively heavy, they fall to the ground within about three months, so their cooling effect is very short-lived. But volcanic debris also includes sulfur dioxide. This gas combines with water vapor and dust in the atmosphere to form sulfate aerosols, which reflect sunlight away from the Earth's surface. These aerosols are lighter than ash particles and can remain in the atmosphere for a year or more. Their cooling effect outweighs the warming caused by volcanic greenhouse gases – the eruption of Mount Pinatubo in 1991 caused a 0.5 °C drop in global temperature. Volcanic eruptions spew out lava, carbon dioxide (CO_2) ash and particles. Although CO_2 has a warming effect, average volcanic CO_2 emissions are less than 1% of emissions from current human activities. Large volumes of gases and ash can influence climatic patterns for years by increasing planetary reflectivity causing atmospheric cooling.

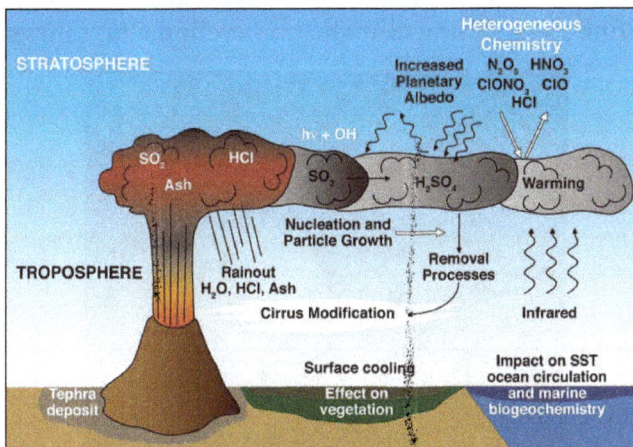

Ocean currents - The oceans are a major component of the climate system. Ocean currents are located at the ocean surface and in deep water below 300 meters (984 feet). They can move water horizontally and vertically and occur on both local and global scales. The ocean has an interconnected current, or circulation, system powered by wind, tides, the Earth's rotation (Coriolis effect), the sun (solar energy), and water density differences. The topography and shape of ocean basins and nearby landmasses also influence ocean currents.

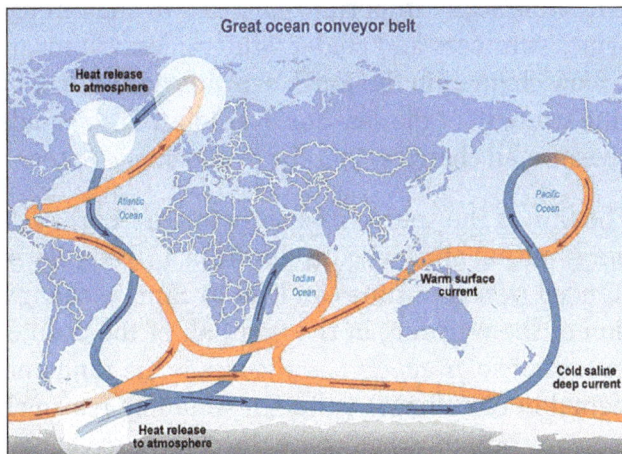

Deep ocean currents are density-driven and differ from surface currents in scale, speed, and energy. Water density is affected by the temperature, salinity (saltiness), and depth of the water. The colder and saltier the ocean water, the denser it is. The greater the density differences between different layers in the water column, the greater the mixing and circulation. Density differences in ocean water contribute to a global-scale circulation system, also called the global conveyor belt.

The global conveyor belt includes both surface and deep ocean currents that circulate the globe in a 1,000-year cycle. This circulation is the result of two simultaneous processes: warm surface currents carrying less dense water away from the Equator toward the poles, and cold deep ocean currents carrying denser water away from the poles toward the Equator. The ocean's global circulation system plays a key role in distributing heat energy, regulating weather and climate, and cycling vital nutrients and gases.

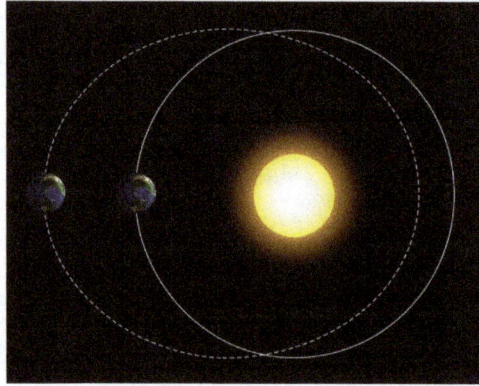

Earth orbital changes - Shifts and wobbles in the Earth's orbit can trigger changes in climate such as the beginning and end of ice ages. The last ice age ended about 12,000 years ago and the next cooling cycle may begin in about 30,000 years. But orbital changes are so gradual they're only noticeable over thousands of years – not decades or centuries. The earth makes one full orbit around the sun each year. It is tilted at an angle of 23.5° to the perpendicular plane of its orbital path. Changes in the tilt of the earth can lead to small but climatically important changes in the strength of the seasons, more tilt means warmer summers and colder winters; less tilt means cooler summers and milder winters. Slow changes in the Earth's orbit lead to small but climatically important changes in the strength of the seasons over tens of thousands of years. Climate feedbacks amplify these small changes, thereby producing ice ages.

Solar variations - The Sun is the source of energy for the Earth's climate system. Although the Sun's energy output appears constant from an everyday point of view, small changes over an extended period of time can lead to climate changes. Some scientists suspect that a portion of the warming in the first half of the 20th century was due to an increase in the output of solar energy. As the sun is the fundamental source of energy that is instrumental in our climate system it would be reasonable to assume that changes in the sun's energy output would cause the climate to change. Scientific studies demonstrate that solar variations have performed a role in past climate changes. For instance a decrease in solar activity was thought to have triggered the Little Ice Age between approximately 1650 and 1850, when Greenland was largely cut off by ice from 1410 to the 1720s and glaciers advanced in the Alps.

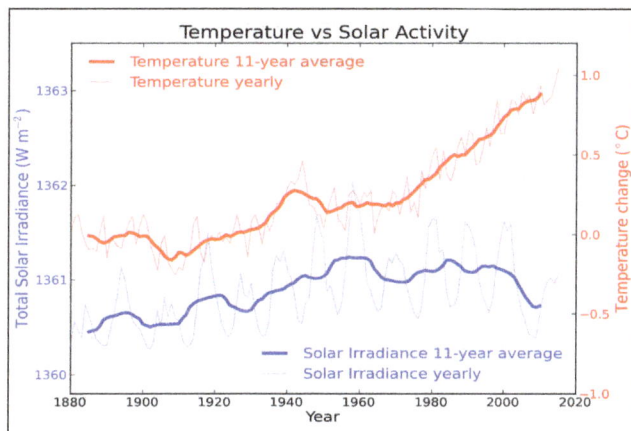

Current global warming however cannot be explained by solar variations. Since 1750, the average amount of energy coming from the Sun either remained constant or increased slightly. If global warming was caused by a more active sun, scientists would expect to see warmer temperatures in all layers of the atmosphere. They have only observed a cooling in the upper atmosphere, a warming at the surface and in the lower parts of the atmosphere. This is due to greenhouse gasses capturing heat in the lower atmosphere. Also climate models that include solar irradiance changes cannot reproduce last century's observed temperature trend without including a rise in greenhouse gases.

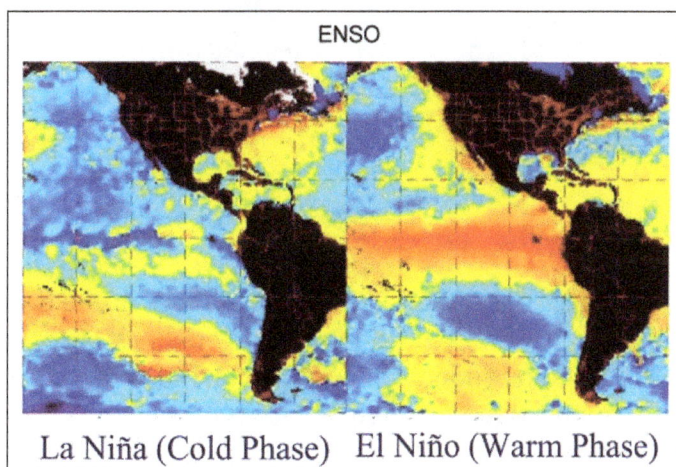

La Niña (Cold Phase)   El Niño (Warm Phase)

Internal variability - Some changes in climate have no external trigger. These changes are instead caused by interactions within the climate system itself, often involving positive feedbacks. One example is the El Niño–La Niña cycle, which can cause temporary warming and cooling. Both phenomena affect atmospheric circulation patterns and influence global climate. While El Niño increases global temperature, La Niña decreases it. This cycle repeats itself on a timescale of about five years. But these changes are short-term, only lasting a few years.

Another example of internal variability is the Arctic oscillation (AO), which is associated with changing patterns of air pressure in the northern hemisphere. This phenomenon brings warmer weather to parts of Europe and North America, leaving the Arctic colder than usual. The other phase of the AO brings the opposite conditions, resulting in a warmer-than-usual Arctic and colder weather in the sub-polar regions. Because of this seesaw effect, the AO has little effect on global temperatures, but can significantly influence local and regional weather.

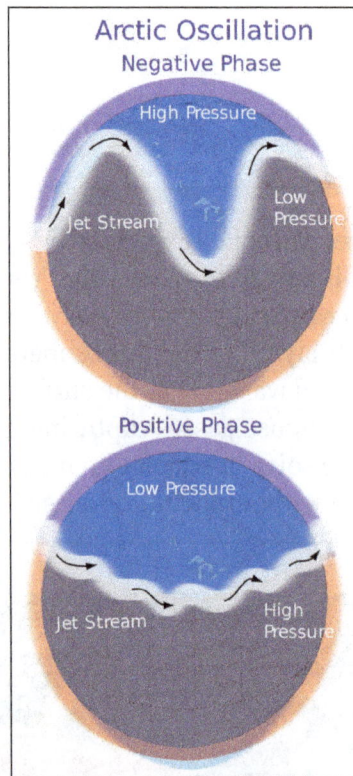

Arctic Oscillation
Negative Phase
High Pressure
Jet Stream
Low Pressure

Positive Phase
Low Pressure
Jet Stream
High Pressure

Many natural factors affect the climate, including changes in the Sun's output or the Earth's orbit, large volcanic eruptions and internal variability such as El Niño. Scientists measure these effects, but none can account for the observed trend since 1970. Scientists can only account for recent global warming by including the effects of human greenhouse gas emissions.

The climate varies on multiple timescales, but now humans are the main agents of change and are likely to remain so for the next few centuries. It is generally understood that human-induced climate change causes global warming, but what is not adequately appreciated are the direct influences on heavy rainfalls, drought, and storms, at great cost to society and the environment. Although the climate change effects are modest, perhaps 5 to 15% for these events, once thresholds are crossed, things break and damage increases nonlinearly. These aspects are not properly factored into costs of climate change, and preparation for expected effects is woefully inadequate, exacerbating damage.

Human activities have led to the release of carbon dioxide and other heat-trapping "greenhouse" gases in sufficient quantity to change the composition of the atmosphere, resulting in an accumulation of heat in the Earth's system, commonly referred to as "global warming". The Earth's climate has responded through higher temperatures in the atmosphere, land and ocean, ice melting, rising sea level, and increases in extreme weather events (heat waves, wildfires, heavy rains and flooding). The calendar year 2016 is by far the warmest on record for the global mean surface temperatures (GMSTs). It easily beat out 2015, which in turn beat out the previous record holder 2014. Meanwhile, 2017 is now ranked third (or second, depending on dataset). There is no doubt whatsoever that the planet is warming and it has major consequences for other aspects of climate. However, there is also considerable natural variability manifested in the GMST record; the biggest fluctuations from year to year are associated with El Niño events. Decadal variations led to a pause in warming from 2000 to 2013. A major El Niño from 2015-16 somewhat inflated the GMST values, and 2017 values dropped slightly, as a result.

How Global Warming affects Extreme Events

Global warming affects the overall heating of the planet mainly from human changes in the atmospheric composition, which leads to general temperature increases in the atmosphere and oceans, and melting ice, there are substantial impacts on extreme events. Indeed, the biggest impacts of climate change on society and the environment arise from changes in extremes. These are realized through the daily weather systems, which naturally produce tremendous variability on all time scales and over many different spatial scales. Hence just by chance, extreme values of temperatures, precipitation, or wind, and so forth, occur from vigorous weather systems. With global warming, some of these extremes are pushed higher and beyond previous values, creating new records. Moreover, global warming often pushes values over various thresholds used for design purposes: whether for heat, rain, wind, or sea level, and accordingly things break. This also means that the events and the new records are episodic. There is not a continuous level of high values, rather the values fluctuate substantially as they have always done with natural weather patterns. It also means that in one month records are broken at one location, while in the next month records break somewhere else, and then somewhere else again. The fact that the extremes occur in different places over time, means that the public often does not connect them to climate change, and their accumulated effects have been greatly underestimated by many. It also means that because of the natural climate variability from year to year, it is often difficult to conclusively detect the climate change influences – an issue of signal-to-noise.

Heat Waves

The most obvious expectation is for an increase in short-duration heat waves and their impacts as overall temperatures rise. Often these result in temperature rises beyond anything previously experienced in recorded history, and this has been borne out in many studies. Heat waves nearly always occur in association with a strong slow-moving anticyclone. The major European heat wave in the summer of 2003 , was one of the first to be well documented both in terms of its detection as extremely unusual, and attribution to anthropogenic climate change using climate models. There were major consequences in terms of wildfires, and loss of life. A more recent example is the extreme Russian heat wave of 2010, again with widespread wildfires, smoke, agricultural losses, and loss of life. Some confusion and debate has occurred in the scientific literature about this event over the cause and rarity of the weather situation, versus the role of human-induced warming. This confusion arises because the weather events (strong anticyclones) tend to occur naturally; while it is the global warming that pushes what would have been an extreme event anyway into one that goes well outside previous bounds and causes major strife.

High temperatures can result in detrimental health, economic, and social impacts. The European 2003 and the Russian 2010 heat waves caused, respectively, almost 70,000 and 55,000 deaths , while an average of 658 deaths were reported annually during 1999-2009 in the United States alone due to excessive heat. Extreme high temperatures may cause human casualties in large cities and have profound impacts on farms due to re-duced crop productivity and adverse effects on animals, including mortality. Temperature extremes stress infrastructure, transportation, water supply, and electricity demand; severely affect ecosystems and forests, and increase wildfire activity. Heat strokes—the most lethal condition of hyperthermia—can be caused by exposure to high ambient environmental temperatures. More frequent, more intense, and longer lasting heat waves are robustly projected in the 21st Century as a result of human-induced global warming.

Drought and Wildfire

In the United States, and indeed in mid-latitude continental areas around the world, there is a strong negative correlation between monthly-mean temperatures and precipitation in the summer half year, as there is year-round in the tropics. Heat waves, especially ones of longer duration, often occur in association with drought. The anticyclonic conditions that persist in a drought situation make for dry settled weather, with little or no precipitation. Under these circumstances, the land and vegetation dry out, and the modest extra heat from global warming exacerbates the dry conditions. Evaporative cooling ceases as plants wilt, wildfire risk increases, and the heat intensifies. That in turn increases the atmospheric demand for moisture, further drying out the vegetation in a vicious cycle.

The warmest year on record for the United States as a whole was 2012 when there was a widespread drought in association with persistent anticyclonic conditions over much of the country. Extreme drought was estimated to cover 39% of the country at its peak in September 2012, rivaling the Dust Bowl years in the early 1930s. According to the Sept. 4, 2012 drought monitor, 64% of the country was in moderate to extreme drought. Wildfires became endemic in many places, and firefighting costs soared. As a result of these events and the agricultural and livestock losses, the net cost has been estimated as over $75B, although a partial accounting by NOAA lists it as $32B. Wildfire Today reports the fire-fighting costs alone in 2012 were $2B.

Perhaps the best example of how climate change can lead to an increase in drought conditions is in the American West, particularly California. A record-setting drought began in 2012 and persisted until 2016 in spite of the big El Niño event (which favors more storms coming into the West Coast). It included the lowest annual precipitation on record, the highest annual temperature, as well as the most extreme drought indicators ever recorded in California. Along with widespread water shortages, the drought brought prolonged and costly wildfires. Indeed, wildfires were rampant throughout the West, especially in the summer of 2015, with wildfires widespread in Alaska, western Canada, Washington, Oregon and California. In May 2016, a major wildfire broke out in Fort McMurray, Alberta following 5 to 8 months of prolonged (El Niño related) drought. Major wildfires continued again in August 2016 and July 2017 in California, and the consensus has become that the wildfire season in California is now almost continuous. In early 2017, in association with unusually high sea temperatures in the subtropical North Pacific, the drought in California was abated with heavy rains and snows, leading to flooding in many areas. This was a boon in terms of snowpack to the Sierra Nevadas and Rocky Mountains.

Certain bugs and diseases flourish under these warmer and dryer conditions, such as the bark beetle, which is decimating forests across the West. Increased carbon dioxide is not good for plants.

Storms and Precipitation

Perhaps less obvious, but even more dangerous than heat, are the effects of a warming planet on the water cycle in which the oceans play a key role. The atmosphere holds about 4% more moisture per 1°F (or 7% per 1°C) increase in temperature, which leads to increased water vapor in the atmosphere, and this provides the biggest influence on precipitation. It is undisputed that water vapor is a powerful greenhouse gas, and hence this amplifies the original warming substantially. In addition, sea surface temperatures have warmed by more than 1°F since the 1970s, and over the oceans this has led to 5 to 10% more water vapor in the atmosphere.

Storms, whether individual thunderstorms, extratropical rain or snow storms, or tropical cyclones and hurricanes, supplied by increased moisture, produce more intense precipitation events, even in places where total precipitation is decreasing. The increased moisture and related latent heat release can intensify storms and perhaps double the original change so that the precipitation increases 5 to 20%. The effect on the storm depends on where the precipitation and released heat occurs relative to the storm center. For hurricanes, the effect is direct and the result can be doubled or more. For extratropical storms the effects are more complicated and the effect is a factor of 1 to 2 and varies from storm to storm.

Nevertheless, it leads to much stronger and more intense rains, and snows, and it increases risk of flooding that exceeds previous bounds for extreme weather events. At the same time, dry spells in between such events also increase. Indeed, in places where it is not raining, the extra heat dries things out, exacerbating heat waves as the evaporative cooling is lost. Hence, droughts set in quicker and become more intense, increasing risk of wildfire. This is especially a dangerous problem in the U.S. West.

However, in Colorado, the unprecedented widespread flooding along the Front Range in September 2013 is a case in point. The moisture sources came from very warm ocean regions to the south (the Gulf of Mexico and especially from west of Mexico) that undoubtedly had a global warming component. More recently, widespread flooding occurred in Missouri, in Houston in April 2016, in Louisiana in August 2016, and in the Carolinas from hurricane Matthew in October 2016. The major winter storm "Jonas" that "bombed" Washington D.C. with several feet of snow in January 2016 is another example of such an extreme event. Meanwhile torrential rains, flooding, mud slides, and loss of life occurred in South America: in northern Chile in late February 2017, in Peru in March, and Colombia in early April in association with a coastal El Niño that led to very high sea temperatures off the Pacific coast in combination with global warming.

Without climate change, these events would have been properly labeled as "1000-year events". However, because of climate change and its effects on the environment, they are no longer 1 in 1000-year events, and instead, they are now more likely 1 in 50-year or 100-year events. They are still uncommon, but not unlikely.

Tropical Storms and Hurricanes

Tropical storms and hurricanes/typhoons mostly occur in the deep tropics in summer in association with high sea surface temperatures (SSTs) over 27 °C. In turn these reflect high ocean heat content (OHC) below the surface and it is this heat energy that is transferred into the atmosphere through evaporation, moistening the atmosphere, while evaporative cooling occurs in the ocean. The fuel for tropical storms and hurricanes comes from the release of the latent heat in heavy rainfall as the moisture is gathered into the storm and condensed.

One harmful aspect of hurricanes is the fierce winds that cause destruction to people's homes and other buildings and infrastructure. However, hurricanes are also responsible for huge storm surges in coastal regions that can be very damaging and are expected to become much worse due to both stronger winds and higher sea levels. The most widespread damage, though, is actually the flooding from torrential rains that can extend hundreds of miles from the coast.

One major source of variability in tropical SSTs is the El Niño phenomenon that produces a warming in the central and eastern Pacific with a corresponding shift in tropical storm activity into that region at the expense of other regions. Hurricanes become more frequent in the eastern North Pacific but decrease in the Atlantic, for example. Indeed, there is always a competition throughout the tropics for where the main activity occurs, and high SSTs are the main factor. Once activity is underway in one region of the tropics, it tends to suppress activity elsewhere by creating a large overturning circulation in the atmosphere that creates subsiding stable air elsewhere and wind-shear in in-between regions (where the low-level winds and upper level winds in the troposphere at jet stream level are indifferent directions and/or speeds), and this tends to blow a developing vortex apart. Accordingly, tropical storms are clustered and cannot occur everywhere at once.

In general, climate warming invigorates tropical storm activity by adding energy to the storms, but it can be manifested in several ways. With climate change, it is expected that hurricanes will contain heavier rains and become more intense, longer lasting, and possibly larger in size, but fewer in number, as one big storm essentially replaces the effects of several smaller weaker storms in terms of the heat energy pulled out of the ocean. Owing to the large natural climate variability from year-to-year and unreliable records prior to the satellite era, it is difficult to clearly detect climate change influences on tropical storm activity. "Detection" relies on a climate signal that is larger than the noise of natural variability, confounded also in this case by unreliable data. So, it is not that there is no signal, but rather that the noise is large. Indeed, there is very compelling evidence that there is a climate signal to increased tropical storm activity.

Examples of increased activity are the record-breaking exceptionally large number and strength of storms in the Atlantic in 2005, super storm Sandy on the East Coast in 2012, the strongest land-falling typhoon on record: Haiyan in 2013 that went through the Philippines, and the very strong storms recorded in several regions in 2015 and 2016 (strongest in the southern Hemisphere – Winston in 2016 that went through Fiji). Then in 2017 it was the Atlantic's turn, with Harvey, Irma and Maria creating devastation in Texas, Florida and the Caribbean Islands, and Puerto Rico. The year 2015 is the most active year globally for hurricanes/typhoons ever. The latter is in part because it was an El Niño year, but it highlights the fact that high sea surface temperatures from whatever reason produce bigger and stronger storms. At the same time, there are quiet years, that highlight the large variability.

Costs of flooding for a number of events have been assigned and hurricanes Katrina, Rita and Wilma in 2005 cost over $180 billion (2011 prices). The recent Atlantic hurricanes in 2017 are estimated to have damages of over $230 billion.

Snowfall and Snow Cover

In winter over the northern hemisphere land, the snow season is getting shorter at each end as more precipitation arrives as rain. Generally, the biggest snowfall occurs with temperatures just below freezing, and hence in mid-winter, the prospects, as observed, are for bigger snowfalls and larger snow pack from November through January. With climate change, it is no longer "too cold to snow" very often. In contrast, snow pack is observed to be much reduced across the northern hemisphere from March through August 1966 to 2014. Due to global warming, snow melt starts sooner, runoff occurs sooner in the spring, and the risk of drought and water shortages are greater in summer, along with wildfire, and insect pest infestations.

Human Activities Contribute to Climate Change

Human activities contribute to climate change by causing changes in Earth's atmosphere in the amounts of greenhouse gases, aerosols (small particles), and cloudiness. The largest known contribution comes from the burning of fossil fuels, which releases carbon dioxide gas to the atmosphere. Greenhouse gases and aerosols affect climate by altering incoming solar radiation and outgoing infrared (thermal) radiation that are part of Earth's energy balance. Changing the atmospheric abundance or properties of these gases and particles can lead to a warming or cooling of the climate system. Since the start of the industrial era (about 1750), the overall effect of human activities on climate has been a warming influence. The human impact on climate during this era greatly exceeds that due to known changes in natural processes, such as solar changes and volcanic eruptions.

Greenhouse Gases

Human activities result in emissions of four principal greenhouse gases: Carbon dioxide (CO_2), methane (CH_4), nitrous oxide (N_2O) and the halocarbons (a group of gases containing fluorine, chlorine and bromine). These gases accumulate in the atmosphere, causing concentrations to increase with time. Significant increases in all of these gases have occurred in the industrial era. All of these increases are attributable to human activities:

- Carbon dioxide has increased from fossil fuel use in transportation, building heating and cooling and the manufacture of cement and other goods. Deforestation releases CO_2 and reduces its uptake by plants. Carbon dioxide is also released in natural processes such as the decay of plant matter.

- Methane has increased as a result of human activities related to agriculture, natural gas distribution and landfills. Methane is also released from natural processes that occur, for example, in wetlands. Methane concentrations are not currently increasing in the atmosphere because growth rates decreased over the last two decades.

- Nitrous oxide is also emitted by human activities such as fertilizer use and fossil fuel burning. Natural processes in soils and the oceans also release N_2O.

- Halocarbon gas concentrations have increased primarily due to human activities. Natural processes are also a small source. Principal halocarbons include the chlorofluorocarbons (e.g. CFC-11 and CFC-12), which were used extensively as refrigeration agents and in other industrial processes before their presence in the atmosphere was found to cause stratospheric ozone depletion. The abundance of chlorofluorocarbon gases is decreasing as a result of international regulations designed to protect the ozone layer.

- Ozone is a greenhouse gas that is continually produced and destroyed in the atmosphere by chemical reactions. In the troposphere, human activities have increased ozone through the release of gases such as carbon monoxide, hydrocarbons and nitrogen oxide, which chemically react to produce ozone. As mentioned above, halocarbons released by human activities destroy ozone in the stratosphere and have caused the ozone hole over Antarctica.

- Water vapour is the most abundant and important greenhouse gas in the atmosphere. However, human activities have only a small direct influence on the amount of atmospheric water vapour. Indirectly, humans have the potential to affect water vapour substantially by changing climate. For example, a warmer atmosphere contains more water vapour. Human activities also influence water vapour through CH_4 emissions, because CH_4 undergoes chemical destruction in the stratosphere, producing a small amount of water vapour.

- Aerosols are small particles present in the atmosphere with widely varying size, concentration and chemical composition. Some aerosols are emitted directly into the atmosphere while others are formed from emitted compounds. Aerosols contain both naturally occurring compounds and those emitted as a result of human activities. Fossil fuel and biomass burning have increased aerosols containing sulphur compounds, organic compounds and black carbon (soot). Human activities such as surface mining and industrial processes have increased dust in the atmosphere. Natural aerosols include mineral dust released from the surface, sea salt aerosols, biogenic emissions from the land and oceans and sulphate and dust aerosols produced by volcanic eruptions.

Radiative Forcing of Factors affected by Human Activities

The contributions to radiative forcing from some of the factors influenced by human activities. The values reflect the total forcing relative to the start of the industrial era. The forcings for all greenhouse gas increases, which are the best understood of those due to human activities, are positive because each gas absorbs outgoing infrared radiation in the atmosphere. Among the greenhouse gases, CO_2 increases have caused the largest forcing over this period. Tropospheric ozone increases have also contributed to warming, while stratospheric ozone decreases have contributed to cooling.

Aerosol particles influence radiative forcing directly through reflection and absorption of solar and infrared radiation in the atmosphere. Some aerosols cause a positive forcing while others cause a negative forcing. The direct radiative forcing summed over all aerosol types is negative. Aerosols also cause a negative radiative forcing indirectly through the changes they cause in cloud properties. Human activities since the industrial era have altered the nature of land cover over the globe, principally through changes in croplands, pastures and forests. They have also modified the reflective properties of ice and snow. Overall, it is likely that more solar radiation is now being reflected from Earth's surface as a result of human activities. This change results in a negative forcing.

Aircraft produce persistent linear trails of condensation ('contrails') in regions that have suitably low temperatures and high humidity. Contrails are a form of cirrus cloud that reflect solar radiation and absorb infrared radiation. Linear contrails from global aircraft operations have increased Earth's cloudiness and are estimated to cause a small positive radiative forcing.

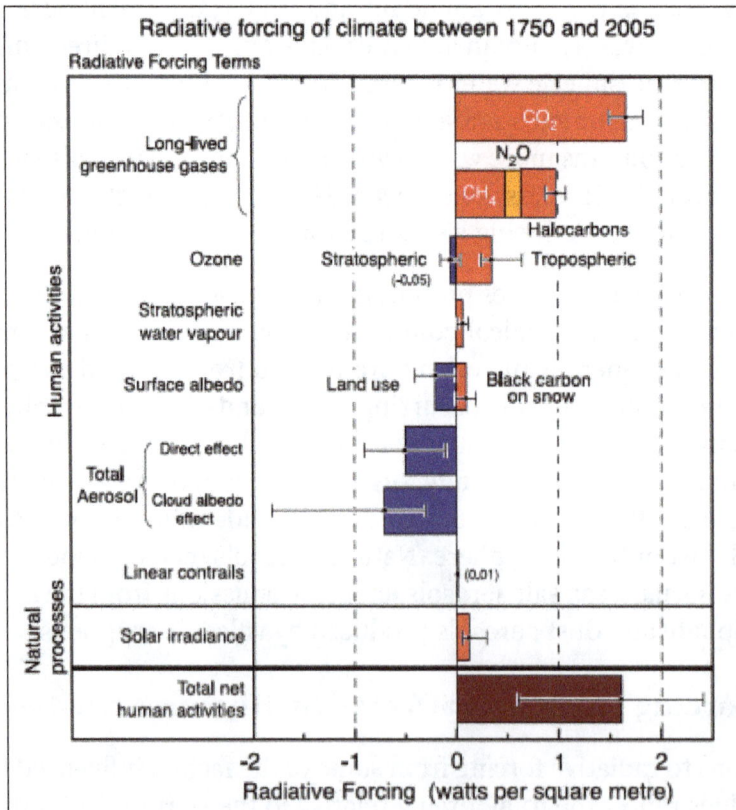

Figure shows the principal components of the radiative forcing of climate change. All these radiative forcings result from one or more factors that affect climate and are associated with human activities or natural processes as discussed in the text. The values represent the forcings in 2005 relative to the start of the industrial era. Human activities cause significant changes in long-lived gases, ozone, water vapour, surface albedo, aerosols and contrails. The only increase in natural forcing of any significance between 1750 and 2005 occurred in solar irradiance. Positive forcings lead to warming of climate and negative forcings lead to a cooling. The thin black line attached to each coloured bar represents the range of uncertainty for the respective value.

Radiative Forcing from Natural Changes

Natural forcings arise due to solar changes and explosive volcanic eruptions. Solar output has increased gradually in the industrial era, causing a small positive radiative forcing. This is in addition to the cyclic changes in solar radiation that follow an 11-year cycle. Solar energy directly heats the climate system and can also affect the atmospheric abundance of some greenhouse gases, such as stratospheric ozone. Explosive volcanic eruptions can create a short-lived (2 to 3 years) negative forcing through the temporary increases that occur in sulphate aerosol in the stratosphere. The stratosphere is currently free of volcanic aerosol, since the last major eruption was in 1991 (Mt. Pinatubo).

The differences in radiative forcing estimates between the present day and the start of the industrial era for solar irradiance changes and volcanoes are both very small compared to the differences in radiative forcing estimated to have resulted from human activities. As a result, in today's atmosphere, the radiative forcing from human activities is much more important for current and future climate change than the estimated radiative forcing from changes in natural processes.

External Force Mechanisms

Orbital Variations

Changes in orbital eccentricity affect the Earth-sun distance. Currently, a difference of only 3 percent (5 million kilometers) exists between closest approach (perihelion), which occurs on or about January 3, and furthest departure (aphelion), which occurs on or about July 4. This difference in distance amounts to about a 6 percent increase in incoming solar radiation (insolation) from July to January. The shape of the Earth's orbit changes from being elliptical (high eccentricity) to being nearly circular (low eccentricity) in a cycle that takes between 90,000 and 100,000 years. When the orbit is highly elliptical, the amount of insolation received at perihelion would be on the order of 20 to 30 percent greater than at aphelion, resulting in a substantially different climate from what we experience today.

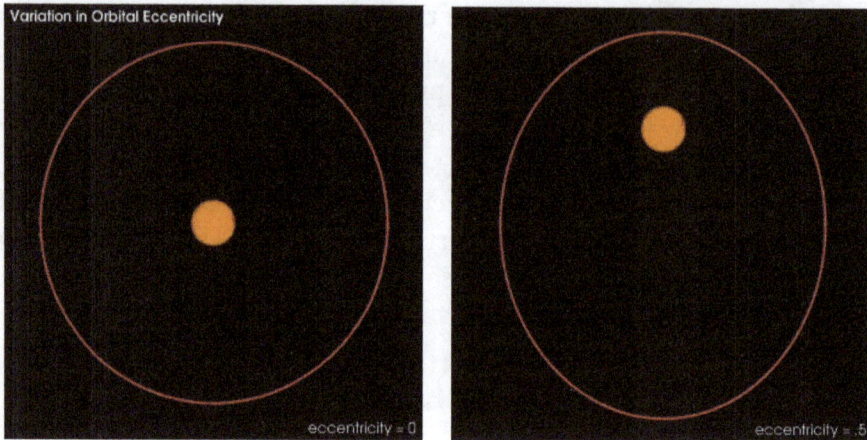

Variation in Orbital Eccentricity

eccentricity = 0

eccentricity = .5

Obliquity (Change in Axial Tilt)

As the axial tilt increases, the seasonal contrast increases so that winters are colder and summers are warmer in both hemispheres. Today, the Earth's axis is tilted 23.5 degrees from the plane of its orbit around the sun. But this tilt changes. During a cycle that averages about 40,000 years, the tilt of the axis varies between 22.1 and 24.5 degrees. Because this tilt changes, the seasons as we know them can become exaggerated. More tilt means more severe seasons—warmer summers and colder winters; less tilt means less severe seasons—cooler summers and milder winters. It's the cool summers that are thought to allow snow and ice to last from year-to-year in high latitudes, eventually building up into massive ice sheets. There are positive feedbacks in the climate system as well, because an Earth covered with more snow reflects more of the sun's energy into space, causing additional cooling.

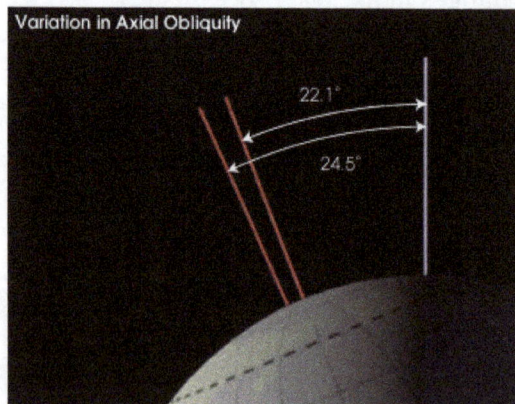

Variation in Axial Obliquity

22.1°

24.5°

Precession

Changes in axial precession alter the dates of perihelion and aphelion, and therefore increase the seasonal contrast in one hemisphere and decrease the seasonal contrast in the other hemisphere.

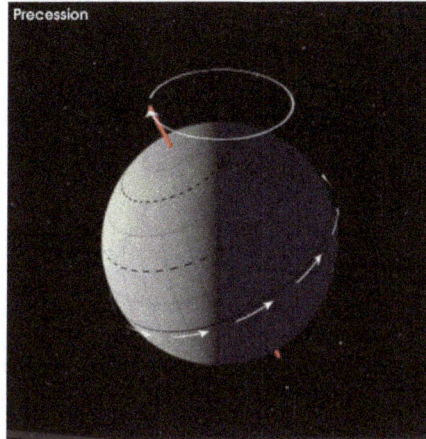

Plate Tectonics

Plate tectonics is a theory dealing with the dynamics of Earth's outer shell—the lithosphere—that revolutionized Earth sciences by providing a uniform context for understanding mountain-building processes, volcanoes, and earthquakes as well as the evolution of Earth's surface and reconstructing its past continents and oceans.

The concept of plate tectonics was formulated in the 1960s. According to the theory, Earth has a rigid outer layer, known as the lithosphere, which is typically about 100 km (60 miles) thick and overlies a plastic (moldable, partially molten) layer called the asthenosphere. The lithosphere is broken up into seven very large continental- and ocean-sized plates, six or seven medium-sized regional plates, and several small ones. These plates move relative to each other, typically at rates of 5 to 10 cm (2 to 4 inches) per year, and interact along their boundaries, where they converge, diverge, or slip past one another. Such interactions are thought to be responsible for most of Earth's seismic and volcanic activity, although earthquakes and volcanoes can occur in plate interiors. Plate motions cause mountains to rise where plates push together, or converge, and continents to fracture and oceans to form where plates pull apart, or diverge. The continents are embedded in the plates and drift passively with them, which over millions of years results in significant changes in Earth's geography.

The theory of plate tectonics is based on a broad synthesis of geologic and geophysical data. It is now almost universally accepted, and its adoption represents a true scientific revolution, analogous in its consequences to quantum mechanics in physics or the discovery of the genetic code in biology. Incorporating the much older idea of continental drift, as well as the concept of seafloor spreading, the theory of plate tectonics has provided an overarching framework in which to describe the past geography of continents and oceans, the processes controlling creation and destruction of landforms, and the evolution of Earth's crust, atmosphere, biosphere, hydrosphere, and climates. During the late 20th and early 21st centuries, it became apparent that plate-tectonic processes profoundly influence the composition of Earth's atmosphere and oceans, serve as a prime cause of long-term climate change, and make significant contributions to the chemical and physical environment in which life evolves.

Principles of Plate Tectonics

In essence, plate-tectonic theory is elegantly simple. Earth's surface layer, 50 to 100 km (30 to 60 miles) thick, is rigid and is composed of a set of large and small plates. Together, these plates constitute the lithosphere. The lithosphere rests on and slides over an underlying partially molten (and thus weaker but generally denser) layer of plastic partially molten rock known as the asthenosphere. Plate movement is possible because the lithosphere-asthenosphere boundary is a zone of detachment. As the lithospheric plates move across Earth's surface, driven by forces as yet not fully understood, they interact along their boundaries, diverging, converging, or slipping past each other. While the interiors of the plates are presumed to remain essentially undeformed, plate boundaries are the sites of many of the principal processes that shape the terrestrial surface, including earthquakes, volcanism, and orogeny (that is, formation of mountain ranges).

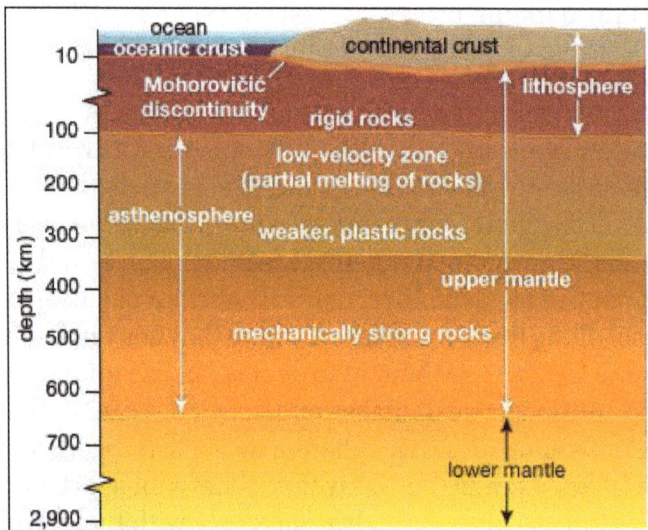

A cross section of Earth's outer layers, from the crust through the lower mantle.

The process of plate tectonics may be driven by convection in Earth's mantle, the pull of heavy old pieces of crust into the mantle, or some combination of both. For a deeper discussion of plate-driving mechanisms.

Earth's Layers

Knowledge of Earth's interior is derived primarily from analysis of the seismic waves that propagate through Earth as a result of earthquakes. Depending on the material they travel through, the waves may either speed up, slow down, bend, or even stop if they cannot penetrate the material they encounter.

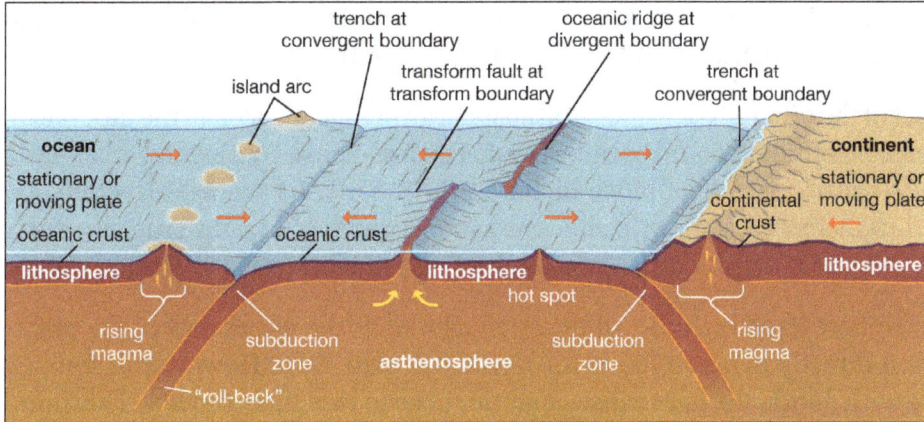

Crustal generation and destruction.

Three-dimensional diagram showing crustal generation and destruction according to the theory of plate tectonics; included are the three kinds of plate boundaries—divergent, convergent (or collision), and strike-slip (or transform).

Collectively, these studies show that Earth can be internally divided into layers on the basis of either gradual or abrupt variations in chemical and physical properties. Chemically, Earth can be divided into three layers. A relatively thin crust, which typically varies from a few kilometres to 40 km (about 25 miles) in thickness, sits on top of the mantle. (In some places, Earth's crust may be up to 70 km [40 miles] thick.) The mantle is much thicker than the crust; it contains 83 percent of Earth's volume and continues to a depth of 2,900 km (1,800 miles). Beneath the mantle is the core, which extends to the centre of Earth, some 6,370 km (nearly 4,000 miles) below the surface. Geologists maintain that the core is made up primarily of metallic iron accompanied by smaller amounts of nickel, cobalt, and lighter elements, such as carbon and sulfur.

There are two types of crust, continental and oceanic, which differ in their composition and thickness. The distribution of these crustal types broadly coincides with the division into continents and ocean basins, although continental shelves, which are submerged, are underlain by continental crust. The continents have a crust that is broadly granitic in composition and, with a density of about 2.7 grams per cubic cm (0.098 pound per cubic inch), is somewhat lighter than oceanic crust, which is basaltic (i.e., richer in iron and magnesium than granite) in composition and has a density of about 2.9 to 3 grams per cubic cm (0.1 to 0.11 pound per cubic inch). Continental crust is typically 40 km (25 miles) thick, while oceanic crust is much thinner, averaging about 6 km (4 miles) in thickness. These crustal rocks both sit on top of the mantle, which is ultramafic in composition (i.e., very rich in magnesium and iron-bearing silicate minerals). The boundary between the crust (continental or oceanic) and the underlying mantle is known as the Mohorovičić discontinuity (also called Moho), which is named for its discoverer, Croatian seismologist Andrija Mohorovičić. The Moho is clearly defined by seismic studies, which detect an acceleration in seismic waves as they pass from the crust into the denser mantle. The boundary between the mantle and the core is also clearly defined by seismic studies, which suggest that the outer part of the core is a liquid.

The effect of the different densities of lithospheric rock can be seen in the different average elevations of continental and oceanic crust. The less-dense continental crust has greater buoyancy, causing it to float much higher in the mantle. Its average elevation above sea level is 840 metres (2,750 feet), while the average depth of oceanic crust is 3,790 metres (12,400 feet). This density difference creates two principal levels of Earth's surface.

The lithosphere itself includes all the crust as well as the upper part of the mantle (i.e., the region directly beneath the Moho), which is also rigid. However, as temperatures increase with depth, the heat causes mantle rocks to lose their rigidity. This process begins at about 100 km (60 miles) below the surface. This change occurs within the mantle and defines the base of the lithosphere and the top of the asthenosphere. This upper portion of the mantle, which is known as the lithospheric mantle, has an average density of about 3.3 grams per cubic cm (0.12 pound per cubic inch). The asthenosphere, which sits directly below the lithospheric mantle, is thought to be slightly denser at 3.4−4.4 grams per cubic cm (0.12−0.16 pound per cubic inch).

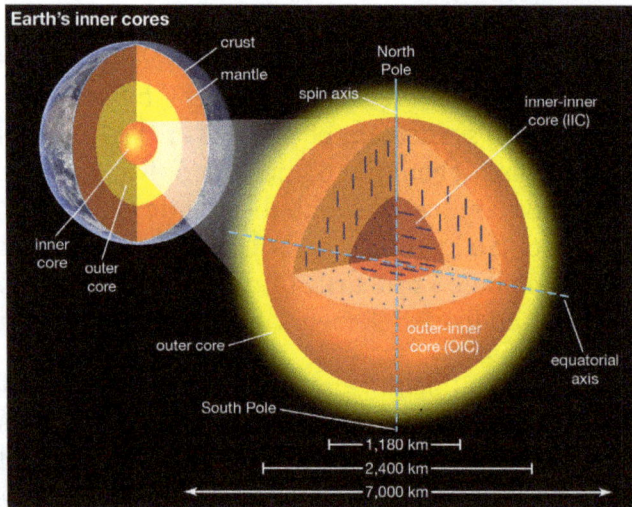

Earth's core: The internal layers of Earth's core, including its two inner cores.

In contrast, the rocks in the asthenosphere are weaker, because they are close to their melting temperatures. As a result, seismic waves slow as they enter the asthenosphere. With increasing depth, however, the greater pressure from the weight of the rocks above causes the mantle to become gradually stronger, and seismic waves increase in velocity, a defining characteristic of the lower mantle. The lower mantle is more or less solid, but the region is also very hot, and thus the rocks can flow very slowly (a process known as creep).

During the late 20th and early 21st centuries, scientific understanding of the deep mantle was greatly enhanced by high-resolution seismological studies combined with numerical modeling and laboratory experiments that mimicked conditions near the core-mantle boundary. Collectively, these studies revealed that the deep mantle is highly heterogeneous and that the layer may play a fundamental role in driving Earth's plates.

At a depth of about 2,900 km (1,800 miles), the lower mantle gives way to Earth's outer core, which is made up of a liquid rich in iron and nickel. At a depth of about 5,100 km (3,200 miles), the outer core transitions to the inner core. Although it has a higher temperature than the outer core, the inner core is solid because of the tremendous pressures that exist near Earth's centre. Earth's inner core is divided into the outer-inner core (OIC) and the inner-inner core (IIC), which differs from one another with respect to the polarity of their iron crystals. The polarity of the iron crystals of the OIC is oriented in a north-south direction, whereas that of the IIC is oriented east-west.

Plate Boundaries

Lithospheric plates are much thicker than oceanic or continental crust. Their boundaries do not usually coincide with those between oceans and continents, and their behaviour is only partly influenced by whether they carry oceans, continents, or both. The Pacific Plate, for example, is entirely oceanic, whereas the North American Plate is capped by continental crust in the west (the North American continent) and by oceanic crust in the east and extends under the Atlantic Ocean as far as the Mid-Atlantic Ridge.

In a simplified example of plate motion shown in the figure, movement of plate A to the left relative to plates B and C results in several types of simultaneous interactions along the plate boundaries. At the rear, plates A and B move apart, or diverge, resulting in extension and the formation of a divergent margin. At the front, plates A and B overlap, or converge, resulting in compression and the formation of a convergent margin. Along the sides, the plates slide past one another, a process called shear. As these zones of shear link other plate boundaries to one another, they are called transform faults.

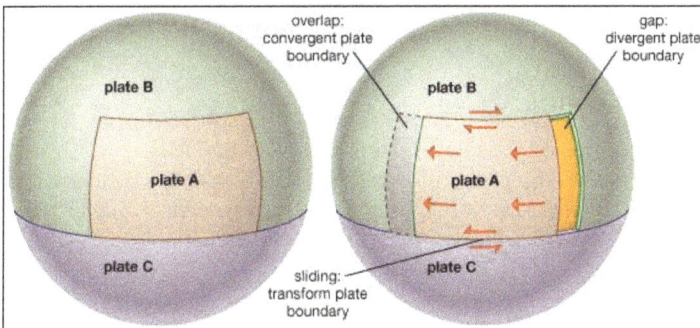

Theoretical diagram showing the effects of an advancing tectonic plate on other adjacent, but stationary, tectonic plates. At the advancing edge of plate A, the overlap with plate B creates a convergent boundary. In contrast, the gap left behind the trailing edge of plate A forms a divergent boundary with plate B. As plate A slides past portions of both plate B and plate C, transform boundaries develop.

Divergent Margins

As plates move apart at a divergent plate boundary, the release of pressure produces partial melting of the underlying mantle. This molten material, known as magma, is basaltic in composition and is buoyant. As a result, it wells up from below and cools close to the surface to generate new crust. Because new crust is formed, divergent margins are also called constructive margins.

Continental Rifting

Upwelling of magma causes the overlying lithosphere to uplift and stretch. (Whether magmatism [the formation of igneous rock from magma] initiates the rifting or whether rifting decompresses the mantle and initiates magmatism is a matter of significant debate.) If the diverging plates are capped by continental crust, fractures develop that are invaded by the ascending magma, prying the continents farther apart. Settling of the continental blocks creates a rift valley, such as the present-day East African Rift Valley. As the rift continues to widen, the continental crust becomes progressively thinner until separation of the plates is achieved and a new ocean is created. The ascending partial melt cools and crystallizes to form new crust. Because the partial melt is basaltic in composition, the new crust is oceanic, and an ocean ridge develops along the site of the former continental rift. Consequently, diverging plate boundaries, even if they originate within continents, eventually come to lie in ocean basins of their own making.

The Thingvellir fracture zone at Thingvellir National Park in southwestern Iceland is an example of a rift valley. The Thingvellir fracture lies in the Mid-Atlantic Ridge, which extends through the centre of Iceland.

Seafloor Spreading

As upwelling of magma continues, the plates continue to diverge, a process known as seafloor spreading. Samples collected from the ocean floor show that the age of oceanic crust increases with distance from the spreading centre—important evidence in favour of this process. These age data also allow the rate of seafloor spreading to be determined, and they show that rates vary from about 0.1 cm (0.04 inch) per year to 17 cm (6.7 inches) per year. Seafloor-spreading rates are much more rapid in the Pacific

Ocean than in the Atlantic and Indian oceans. At spreading rates of about 15 cm (6 inches) per year, the entire crust beneath the Pacific Ocean (about 15,000 km [9,300 miles] wide) could be produced in 100 million years.

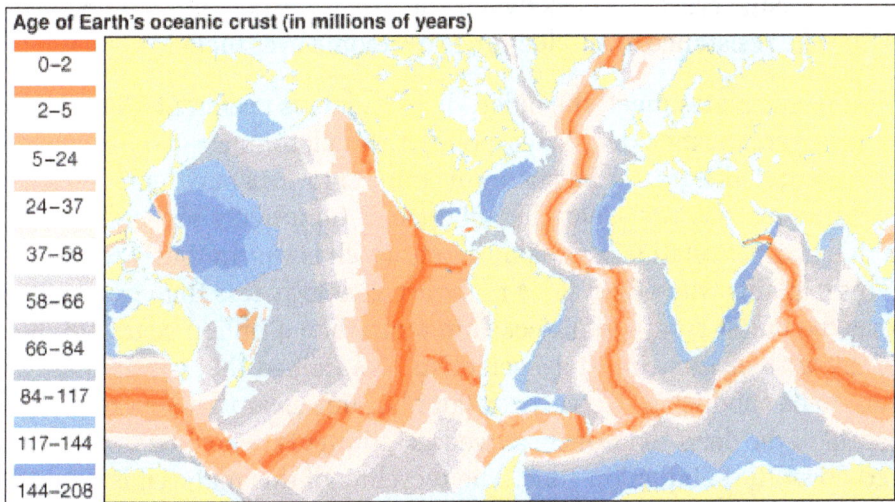

Map showing the age of Earth's oceanic crust and the pattern of seafloor spreading at the global scale.

Divergence and creation of oceanic crust are accompanied by much volcanic activity and by many shallow earthquakes as the crust repeatedly rifts, heals, and rifts again. Brittle earthquake-prone rocks occur only in the shallow crust. Deep earthquakes, in contrast, occur less frequently, due to the high heat flow in the mantle rock. These regions of oceanic crust are swollen with heat and so are elevated by 2 to 3 km (1.2 to 1.9 miles) above the surrounding seafloor. The elevated topography results in a feedback scenario in which the resulting gravitational force pushes the crust apart, allowing new magma to well up from below, which in turn sustains the elevated topography. Its summits are typically 1 to 5 km (0.6 to 3.1 miles) below the ocean surface. On a global scale, these ridges form an interconnected system of undersea "mountains" that are about 65,000 km (40,000 miles) in length and are called oceanic ridges.

Convergent Margins

Given that Earth is constant in volume, the continuous formation of Earth's new crust produces an excess that must be balanced by destruction of crust elsewhere. This is accomplished at convergent plate boundaries, also known as destructive plate boundaries, where one plate descends at an angle—that is, is subducted—beneath the other.

Because oceanic crust cools as it ages, it eventually becomes denser than the underlying asthenosphere, and so it has a tendency to subduct, or dive under, adjacent continental plates or younger sections of oceanic crust. The life span of the oceanic crust is prolonged by its rigidity, but eventually this resistance is overcome. Experiments show that the subducted oceanic lithosphere is denser than the surrounding mantle to a depth of at least 600 km (about 400 miles).

The mechanisms responsible for initiating subduction zones are controversial. During the late 20th and early 21st centuries, evidence emerged supporting the notion that subduction zones preferentially initiate along preexisting fractures (such as transform faults) in the oceanic crust. Irrespective of the exact mechanism, the geologic record indicates that the resistance to subduction is overcome eventually.

Where two oceanic plates meet, the older, denser plate is preferentially subducted beneath the younger, warmer one. Where one of the plate margins is oceanic and the other is continental, the greater buoyancy of continental crust prevents it from sinking, and the oceanic plate is preferentially subducted. Continents are preferentially preserved in this manner relative to oceanic crust, which is continuously recycled into the mantle. This explains why ocean floor rocks are generally less than 200 million years old whereas the oldest continental rocks are more than 4 billion years old. Before the middle of the 20th century, most geoscientists maintained that continental crust was too buoyant to be subducted. However, it later became clear that slivers of continental crust adjacent to the deep-sea trench, as well as sediments deposited in the trench, may be dragged down the subduction zone. The recycling of this material is detected in the chemistry of volcanoes that erupt above the subduction zone.

Two plates carrying continental crust collide when the oceanic lithosphere between them has been eliminated. Eventually, subduction ceases and towering mountain ranges, such as the Himalayas, are created.

Because the plates form an integrated system, it is not necessary that new crust formed at any given divergent boundary be completely compensated at the nearest subduction zone, as long as the total amount of crust generated equals that destroyed.

Subduction Zones

The subduction process involves the descent into the mantle of a slab of cold hydrated oceanic lithosphere about 100 km (60 miles) thick that carries a relatively thin cap of oceanic sediments. The path of descent is defined by numerous earthquakes along a plane that is typically inclined between 30° and 60° into the mantle and is called the Wadati-Benioff zone, for Japanese seismologist Kiyoo Wadati and American seismologist Hugo Benioff, who pioneered its study. Between 10 and 20 percent of the subduction zones that dominate the circum-Pacific ocean basin are subhorizontal (that is, they subduct at angles between 0° and 20°). The factors that govern the dip of the subduction zone are not fully understood, but they probably include the age and thickness of the subducting oceanic lithosphere and the rate of plate convergence.

Most, but not all, earthquakes in this planar dipping zone result from compression, and the seismic activity extends 300 to 700 km (200 to 400 miles) below the surface, implying that the subducted crust retains some rigidity to this depth. At greater depths the subducted plate is partially recycled into the mantle.

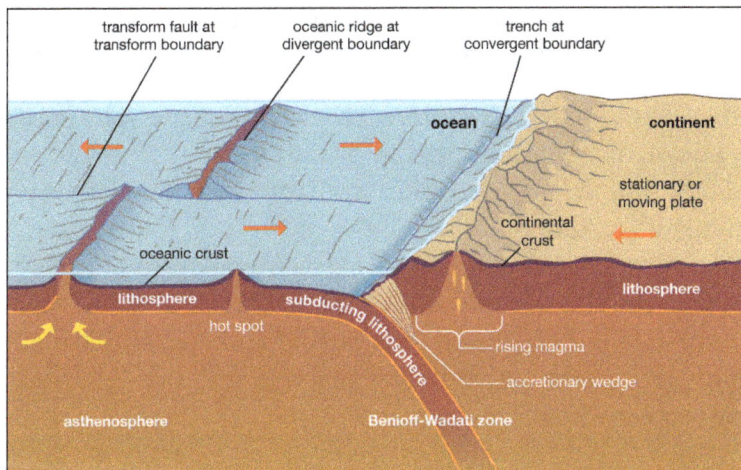

A subducting plate's path (called the Benioff-Wadati [or Wadati-Benioff] zone) is defined by numerous earthquakes along a plane that is typically inclined between 30° and 60° into the mantle.

The site of subduction is marked by a deep trench, between 5 and 11 km (3 and 7 miles) deep, that is produced by frictional drag between the plates as the descending plate bends before it subducts. The overriding plate scrapes sediments and elevated portions of ocean floor off the upper crust of the lower plate, creating a zone of highly deformed rocks within the trench that becomes attached, or accreted, to the overriding plate. This chaotic mixture is known as an accretionary wedge.

The rocks in the subduction zone experience high pressures but relatively low temperatures, an effect of the descent of the cold oceanic slab. Under these conditions the rocks recrystallize, or metamorphose, to form a suite of rocks known as blueschists, named for the diagnostic blue mineral called glaucophane, which is stable only at the high pressures and low temperatures found in subduction zones. At deeper levels in the subduction zone (that is, greater than 30–35 km [about 19–22 miles]), eclogites, which consist of high-pressure minerals such as red garnet (pyrope) and omphacite (pyroxene), form. The formation of eclogite from blueschist is accompanied by a significant increase in density and has been recognized as an important additional factor that facilitates the subduction process.

Island Arcs

When the downward-moving slab reaches a depth of about 100 km (60 miles), it gets sufficiently warm to drive off its most volatile components, thereby stimulating partial melting of mantle in the plate above the subduction zone (known as the mantle wedge). Melting in the mantle wedge produces magma, which is predominantly basaltic in composition. This magma rises to the surface and gives birth to a line of volcanoes in the overriding plate, known as a volcanic arc, typically a few hundred kilometres behind the oceanic trench. The distance between the trench and the arc, known as the arc-trench gap, depends on the angle of subduction. Steeper subduction zones have relatively narrow arc-trench gaps. A basin may form within this region, known as a fore-arc basin, and may be filled with sediments derived from the volcanic arc or with remains of oceanic crust.

If both plates are oceanic, as in the western Pacific Ocean, the volcanoes form a curved line of islands, known as an island arc, that is parallel to the trench, as in the case of the Mariana Islands and the adjacent Mariana Trench. If one plate is continental, the volcanoes form inland, as they do in the Andes of western South America. Though the process of magma generation is similar, the ascending magma may change its composition as it rises through the thick lid of continental crust, or it may provide sufficient heat to melt the crust. In either case, the composition of the volcanic mountains formed tends to be more silicon-rich and iron and magnesium-poor relative to the volcanic rocks produced by ocean-ocean convergence.

Back-arc Basins

Where both converging plates are oceanic, the margin of the older oceanic crust will be subducted because older oceanic crust is colder and therefore more dense. As the dense slab collapses into the asthenosphere, however, it also may "roll back" oceanward and cause extension in the overlying plate. This results in a process known as back-arc spreading, in which a basin opens up behind the island arc. The crust behind the arc becomes progressively thinner, and the decompression of the underlying mantle causes the crust to melt, initiating seafloor-spreading processes, such as melting and the production of basalt; these processes are similar to those that occur at ocean ridges. The geochemistry of the basalts produced at back-arc basins superficially resembles that of basalts produced at ocean ridges, but subtle trace element analyses can detect the influence of a nearby subducted slab.

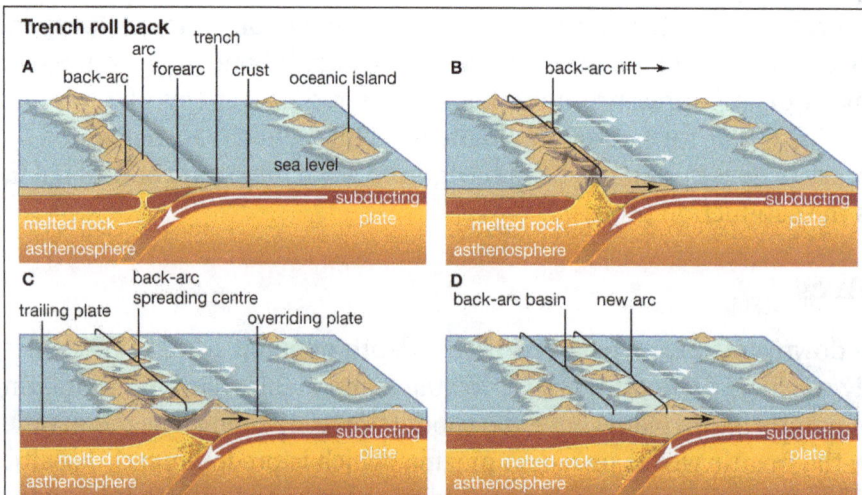

The trench "roll back" process of back-arc basin formation.

This style of subduction predominates in the western Pacific Ocean, in which a number of back-arc basins separate several island arcs from Asia. Examples include the Mariana

Islands, the Kuril Islands, and the main islands of Japan. However, if the rate of convergence increases or if anomalously thick oceanic crust (possibly caused by rising mantle plume activity) is conveyed into the subduction zone, the slab may flatten. Such flattening causes the back-arc basin to close, resulting in deformation, metamorphism, and even melting of the strata deposited in the basin.

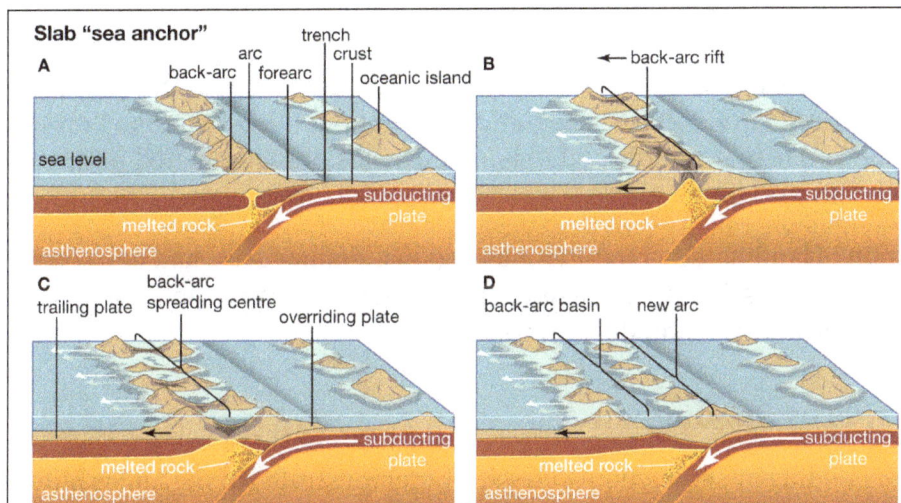

The slab "sea anchor" process of back-arc basin formation.

Mountain Building

If the rate of subduction in an ocean basin exceeds the rate at which the crust is formed at oceanic ridges, a convergent margin forms as the ocean initially contracts. This process can lead to collision between the approaching continents, which eventually terminates subduction. Mountain building can occur in a number of ways at a convergent margin: mountains may rise as a consequence of the subduction process itself, by the accretion of small crustal fragments (which, along with linear island chains and oceanic ridges, are known as terranes), or by the collision of two large continents.

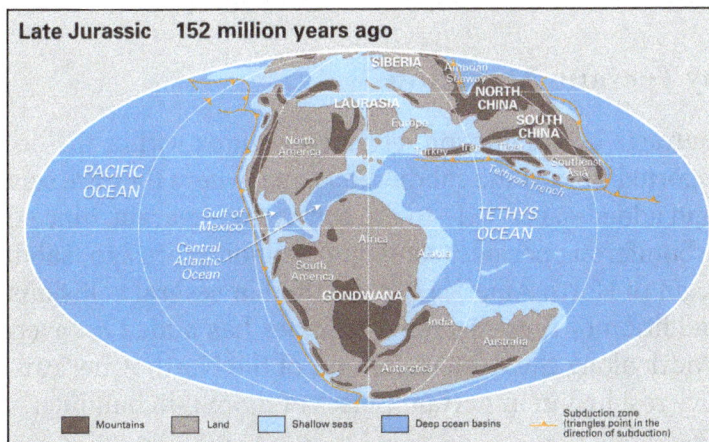

Jurassic paleogeography.

Many mountain belts were developed by a combination of these processes. For example, the Cordilleran mountain belt of North America—which includes the Rocky Mountains as well as the Cascades, the Sierra Nevada, and other mountain ranges near the Pacific coast—developed by a combination of subduction and terrane accretion. As continental collisions are usually preceded by a long history of subduction and terrane accretion, many mountain belts record all three processes. Over the past 70 million years the subduction of the Neo-Tethys Sea, a wedge-shaped body of water that was located between Gondwana and Laurasia, led to the accretion of terranes along the margins of Laurasia, followed by continental collisions beginning about 30 million years ago between Africa and Europe and between India and Asia. These collisions culminated in the formation of the Alps and the Himalayas.

Distribution of landmasses, mountainous regions, shallow seas, and deep ocean basins during the late Jurassic Period. Included in the paleogeographic reconstruction are the locations of the interval's subduction zones.

Mountains by Subduction

Mountain building by subduction is classically demonstrated in the Andes Mountains of South America. Subduction results in voluminous magmatism in the mantle and crust overlying the subduction zone, and, therefore, the rocks in this region are warm and weak. Although subduction is a long-term process, the uplift that results in mountains tends to occur in discrete episodes and may reflect intervals of stronger plate convergence that squeezes the thermally weakened crust upward. For example, rapid uplift of the Andes approximately 25 million years ago is evidenced by a reversal in the flow of the Amazon River from its ancestral path toward the Pacific Ocean to its modern path, which empties into the Atlantic Ocean.

In addition, models have indicated that the episodic opening and closing of back-arc basins have been the major factors in mountain-building processes, which have influenced the plate-tectonic evolution of the western Pacific for at least the past 500 million years.

Mountains by Terrane Accretion

As the ocean contracts by subduction, elevated regions within the ocean basin—terranes—are transported toward the subduction zone, where they are scraped off the descending plate and added—accreted—to the continental margin. Since the late Devonian and early Carboniferous periods, some 360 million years ago, subduction beneath the western margin of North America has resulted in several collisions with terranes. The piecemeal addition of these accreted terranes has added an average of 600 km (400 miles) in width along the western margin of the North American continent, and the collisions have resulted in important pulses of mountain building.

The more gradual transition to the abyssal plain is a sediment-filled region called the continental rise. The continental shelf, slope, and rise are collectively called the continental margin.

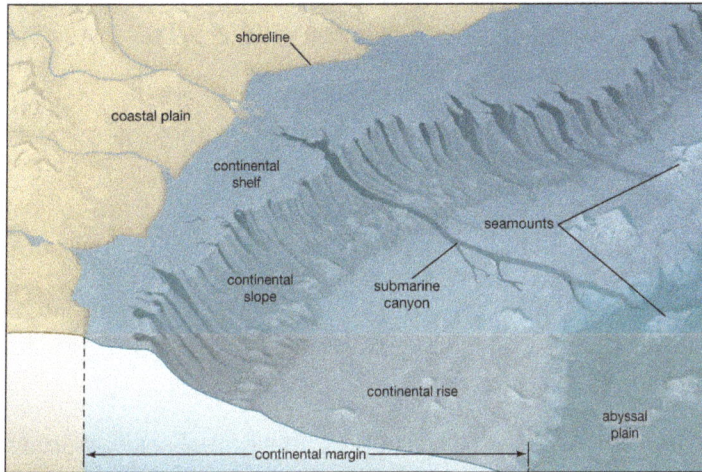

The broad, gentle pitch of the continental shelf gives
way to the relatively steep continental slope.

During these accretionary events, small sections of the oceanic crust may break away from the subducting slab as it descends. Instead of being subducted, these slices are thrust over the overriding plate and are said to be obducted. Where this occurs, rare slices of ocean crust, known as ophiolites, are preserved on land. They provide a valuable natural laboratory for studying the composition and character of the oceanic crust and the mechanisms of their emplacement and preservation on land. A classic example is the Coast Range ophiolite of California, which is one of the most extensive ophiolite terranes in North America. These ophiolite deposits run from the Klamath Mountains in northern California southward to the Diablo Range in central California. This oceanic crust likely formed during the middle of the Jurassic Period, roughly 170 million years ago, in an extensional regime within either a back-arc or a forearc basin. In the late Mesozoic, it was accreted to the western North American continental margin.

Because preservation of oceanic crust is rare, the recognition of ophiolite complexes is very important in tectonic analyses. Until the mid-1980s, ophiolites were thought to represent vestiges of the main oceanic tract, but geochemical analyses have clearly indicated that most ophiolites form near volcanic arcs, such as in back-arc basins characterized by subduction roll-back (the collapse of the subducting plate that causes the extension of the overlying plate). The recognition of ophiolite complexes is very important in tectonic analysis, because they provide insights into the generation of magmatism in oceanic domains, as well as their complex relationships with subduction processes.

Mountains by Continental Collision

Continental collision involves the forced convergence of two buoyant plate margins that results in neither continent being subducted to any appreciable extent. A complex sequence of events ensues that compels one continent to override the other. These processes result in crustal thickening and intense deformation that forces the crust skyward to form huge mountains with crustal roots that extend as deep as 80 km (about 50 miles) relative to Earth's surface, in accordance with the principles of isostasy.

The subducted slab still has a tendency to sink and may become detached and founder (submerge) into the mantle. The crustal root undergoes metamorphic reactions that result in a significant increase in density and may cause the root to also founder into the mantle. Both processes result in a significant injection of heat from the compensatory upwelling of asthenosphere, which is an important contribution to the rise of the mountains.

Continental collisions produce lofty landlocked mountain ranges such as the Himalayas. Much later, after these ranges have been largely leveled by erosion, it is possible that the original contact, or suture, may be exposed.

The balance between creation and destruction on a global scale is demonstrated by the expansion of the Atlantic Ocean by seafloor spreading over the past 200 million years, compensated by the contraction of the Pacific Ocean, and the consumption of an entire ocean between India and Asia (the Tethys Sea). The northward migration of India led to collision with Asia some 40 million years ago. Since that time India has advanced a further 2,000 km (1,250 miles) beneath Asia, pushing up the Himalayas and forming the Plateau of Tibet. Pinned against stable Siberia, China and Indochina were pushed sideways, resulting in strong seismic activity thousands of kilometres from the site of the continental collision.

Transform Faults

Section of the San Andreas Fault in the Carrizo Plain, western California.

Along the third type of plate boundary, two plates move laterally and pass each other along giant fractures in Earth's crust. Transform faults are so named because they are linked to other types of plate boundaries. The majority of transform faults link the offset segments of oceanic ridges. However, transform faults also occur between plate margins with continental crust—for example, the San Andreas Fault in California and the North Anatolian fault system in Turkey. These boundaries are conservative because plate interaction occurs without creating or destroying crust. Because the only motion along these faults is the sliding of plates past each other, the horizontal direction along the fault surface must parallel the direction of plate motion. The fault surfaces are rarely smooth, and pressure may build up when the plates on either side temporarily lock. This buildup of stress may be suddenly released in the form of an earthquake.

Many transform faults in the Atlantic Ocean are the continuation of major faults in adjacent continents, which suggests that the orientation of these faults might be inherited from preexisting weaknesses in continental crust during the earliest stages of the development of oceanic crust. On the other hand, transform faults may themselves be reactivated, and recent geodynamic models suggest that they are favourable environments for the initiation of subduction zones.

Hotspots

Although most of Earth's volcanic activity is concentrated along or adjacent to plate boundaries, there are some important exceptions in which this activity occurs within plates. Linear chains of islands, thousands of kilometres in length, that occur far from plate boundaries are the most notable examples. These island chains record a typical sequence of decreasing elevation along the chain, from volcanic island to fringing reef to atoll and finally to submerged seamount. An active volcano usually exists at one end of an island chain, with progressively older extinct volcanoes occurring along the rest of the chain. Canadian geophysicist J. Tuzo Wilson and American geophysicist W. Jason Morgan explained such topographic features as the result of hotspots.

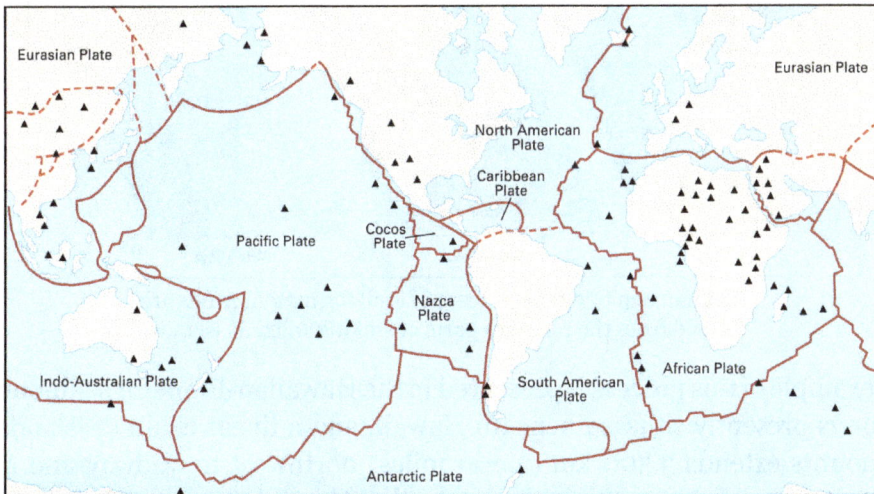

The principal tectonic plates that make up Earth's lithosphere: Also located are several dozen hot spots where plumes of hot mantle material are upwelling beneath the plates.

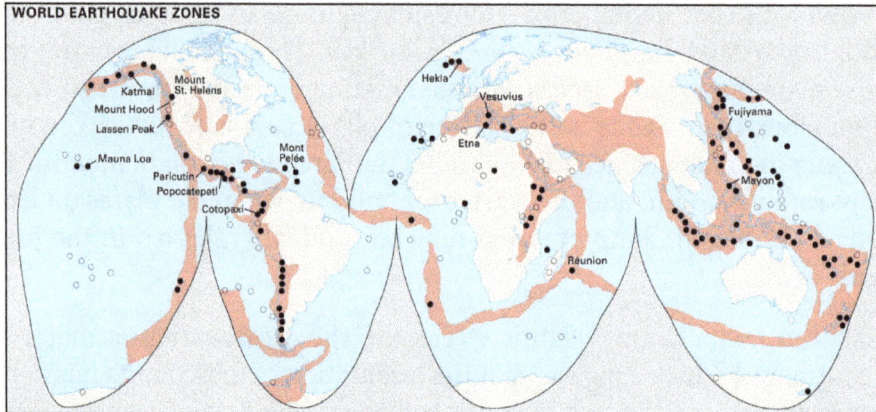

The world's earthquake zones occur in red bands and largely coincide with the boundaries of Earth's tectonic plates. Black dots indicate active volcanoes, whereas open dots indicate inactive ones.

The number of these hotspots is uncertain (estimates range from 20 to 120), but most occur within a plate rather than at a plate boundary. Hotspots are thought to be the surface expression of giant plumes of heat, termed mantle plumes that ascend from deep within the mantle, possibly from the core-mantle boundary, some 2,900 km (1,800 miles) below the surface. These plumes are thought to be stationary relative to the lithospheric plates that move over them. A volcano builds upon the surface of a plate directly above the plume. As the plate moves on, however, the volcano is separated from its underlying magma source and becomes extinct. Extinct volcanoes are eroded as they cool and subside to form fringing reefs and atolls, and eventually they sink below the surface of the sea to form a seamount. At the same time, a new active volcano forms directly above the mantle plume.

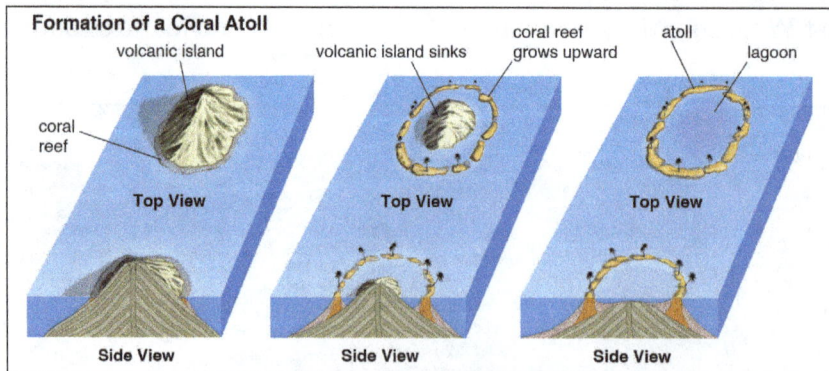

Diagram depicting the process of atoll formation: Atolls are formed from the remnant parts of sinking volcanic islands.

The best example of this process is preserved in the Hawaiian-Emperor seamount chain. The plume is presently situated beneath Hawaii, and a linear chain of islands, atolls, and seamounts extends 3,500 km (2,200 miles) northwest to Midway and a further 2,500 km (1,500 miles) north-northwest to the Aleutian Trench. The age at which volcanism became extinct along this chain gets progressively older with increasing distance from Hawaii—critical evidence that supports this theory. Hotspot volcanism is not restricted to the ocean basins; it also occurs within continents, as in the case of Yellowstone National Park in western North America.

Measurements suggest that hotspots may move relative to one another, a situation not predicted by the classical model, which describes the movement of lithospheric plates over stationary mantle plumes. This has led to challenges to this classic model. Furthermore, the relationship between hotspots and plumes is hotly debated. Proponents of the classical model maintain that these discrepancies are due to the effects of mantle circulation as the plumes ascend, a process called the mantle wind. Data from alternative models suggest that many plumes are not deep-rooted. Instead, they provide evidence that many mantle plumes occur as linear chains that inject magma into fractures, result from relatively shallow processes such as the localized presence of water-rich mantle, stem from the insulating properties of continental crust (which leads to the buildup of trapped mantle heat and decompression of the crust), or are due to instabilities in the interface between continental and oceanic crust. In addition, some geologists note that many geologic processes that others attribute to the behaviour of mantle plumes may be explained by other forces.

Plate Motion

Euler's Contributions

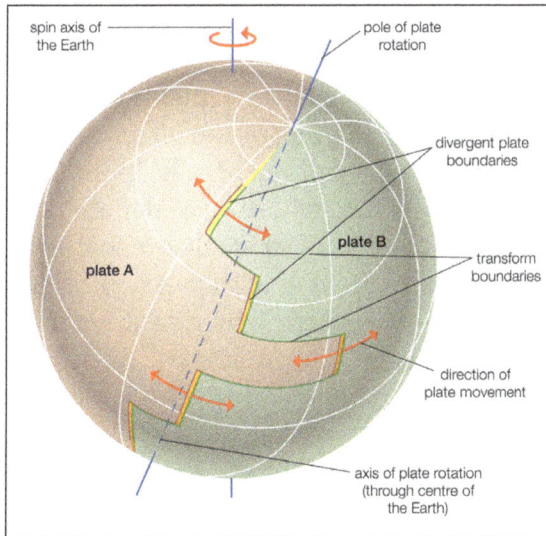

In the 18th century, Swiss mathematician Leonhard Euler showed that the movement of a rigid body across the surface of a sphere can be described as a rotation (or turning) around an axis that goes through the centre of the sphere, known as the axis of rotation. The location of this axis bears no relationship to Earth's spin axis. The point of emergence of the axis through the surface of the sphere is known as the pole of rotation. This theorem of spherical geometry provides an elegant way to define the motion of the lithospheric plates across Earth's surface. Therefore, the relative motion of two rigid plates may be described as rotations around a common axis, known as the axis of spreading. Application of the theorem requires that the plates not be internally deformed—a requirement not absolutely adhered to but one that appears to be a reasonable approximation of what actually happens. Application of this theorem permits the mathematical reconstruction of past plate configurations.

Theoretical depiction of the movement of tectonic plates across Earth's surface: Movement on a sphere of two plates, A and B, can be described as a rotation around a common pole. Circles around that pole correspond to the orientation of transform faults.

Leonhard Euler.

Because all plates form a closed system, all movements can be defined by dealing with them two at a time. The joint pole of rotation of two plates can be determined from their transform boundaries, which are by definition parallel to the direction of motion. Thus, the plates move along transform faults, whose trace defines circles of latitude perpendicular to the axis of spreading, and so form small circles around the pole of rotation. A geometric necessity of this theorem—that lines perpendicular to the transform faults converge on the pole of rotation—is confirmed by measurements. According to this theorem, the rate of plate motion should be slowest near the pole of rotation and increase progressively to a maximum rate along fractures with a 90° angle to it. This relationship is also confirmed by accurate measurements of seafloor-spreading rates.

Past Plate Movements

Plate tectonics involves the movements of Earth's lithospheric plates relative to one another over the planet's weak asthenosphere. This activity changes the positions of all plates with respect to Earth's spin axis and the Equator. To determine the true geographic positions of the plates in the past, investigators have to define their motions, not only relative to each other but also relative to this independent frame of reference. Hotspots, as classically interpreted, provide an example of such a reference frame, assuming they are the sources of plumes that originate within the deep mantle and have relatively fixed positions over time. If this assumption is valid, the motion of the lithosphere above these plumes can be deduced. The hotspot island chains serve this purpose, their trends providing the direction of motion of a plate. The speed of the plate can be inferred from the increase in age of the volcanoes along the chain relative to the distance between the islands.

Earth scientists are able to accurately reconstruct the positions and movements of plates for the past 150 million to 200 million years because they have the oceanic crust record to provide them with plate speeds and direction of movement. However, since older oceanic crust is continuously consumed to make room for new crust, this kind of evidence is not available for earlier intervals of geologic time, making it necessary for investigators to turn to other, less-precise techniques.

Volcanism

Volcanic eruptions are responsible for releasing molten rock, or lava, from deep within the Earth, forming new rock on the Earth's surface. But eruptions also impact the atmosphere.

The gases and dust particles thrown into the atmosphere during volcanic eruptions have influences on climate. Most of the particles spewed from volcanoes cool the planet by shading incoming solar radiation. The cooling effect can last for months to years depending on the characteristics of the eruption. Volcanoes have also caused global warming over millions of years during times in Earth's history when extreme amounts of volcanism occurred, releasing greenhouse gases into the atmosphere.

Even though volcanoes are in specific places on Earth, their effects can be more widely distributed as gases, dust, and ash get into the atmosphere. Because of atmospheric circulation patterns, eruptions in the tropics can have an effect on the climate in both hemispheres while eruptions at mid or high latitudes only have impact the hemisphere they are within.

Below is an overview of materials that make their way from volcanic eruptions into the atmosphere: Particles of dust and ash, sulfur dioxide, and greenhouse gases like water vapor and carbon dioxide.

A huge cloud of volcanic ash and gas rises above Mount Pinatubo, Philippines. Three days later, the volcano exploded in the second-largest volcanic eruption on Earth in the 20th century.

Particles of Dust and Ash

Volcanic ash or dust released into the atmosphere during an eruption shade sunlight and cause temporary cooling. Larger particles of ash have little effect because they fall out of the air quickly. Small ash particles form a dark cloud in the troposphere that shades and cools the area directly below. Most of these particles fall out of the atmosphere within rain a few hours or days after an eruption. But the smallest particles of dust get into the stratosphere and are able to travel vast distances, often worldwide. These tiny particles are so light that they can stay in the stratosphere for months, blocking sunlight and causing cooling over large areas of the Earth.

Sulfur

Often, erupting volcanoes emit sulfur dioxide into the atmosphere. Sulfur dioxide is much more effective than ash particles at cooling the climate. The sulfur dioxide moves into the stratosphere and combines with water to form sulfuric acid aerosols. The sulfuric acid makes a haze of tiny droplets in the stratosphere that reflects incoming solar radiation, causing cooling of the Earth's surface. The aerosols can stay in the stratosphere for up to three years, moved around by winds and causing significant cooling worldwide. Eventually, the droplets grow large enough to fall to Earth.

Greenhouse Gases

Volcanoes also release large amounts of greenhouse gases such as water vapor and carbon dioxide. The amounts put into the atmosphere from a large eruption doesn't change the global amounts of these gases very much. However, there have been times during Earth history when intense volcanism has significantly increased the amount of carbon dioxide in the atmosphere and caused global warming.

References

- What-are-the-effects-of-climate-change: myclimate.org, Retrieved 10 March, 2019
- Natural-climate-change: esrl.noaa.gov, Retrieved 12 May, 2019
- ClimateChange: ucar.edu, Retrieved 8 January, 2019
- Milankovitch: nasa.gov, Retrieved 20 July, 2019
- Plate-tectonics, science: britannica.com, Retrieved 4 April, 2019
- The-characteristics-of-the-eruption, gases-and-dust-particles, how-volcanoes-influence-climate: ucar.edu, Retrieved 2 February, 2019

Chapter 3

Climate Systems and Climate Cycles

Climatic system refers to the system whose components interact and give rise to Earth's climate. The recurring cyclical oscillations within regional or global climate are known as climate cycles. The topics elaborated in this chapter will help in gaining a better perspective about climate systems as well as the different climate cycles such as hydrological cycle and sulfur cycle.

Climate Systems and Change

Controls of Climate

Although almost anything can happen with the weather, climate is more predictable. The weather on a particular winter day in San Diego may be colder than on the same day in Lake Tahoe, but, on average, Tahoe's winter climate is significantly colder than San Diego's. Climate then is the long-term average of weather. Good climate is why we choose to vacation in Hawaii in February, even though the weather is not guaranteed to be good.

Climate is the average of weather in that location over a long period of time, usually for at least 30 years. A location's climate can be described by its air temperature, humidity, wind speed and direction, and the type, quantity, and frequency of precipitation. Climate can change, but only over long periods of time. The climate of a region depends on its position relative to many things.

Latitude

The main factor influencing the climate of a region is latitude because different latitudes receive different amounts of solar radiation.

- The equator receives the most solar radiation. Days are equally long year-round and the sun is just about directly overhead at midday.

- The polar regions receive the least solar radiation. The night lasts six months during the winter. Even in summer, the sun never rises very high in the sky. Sunlight filters through a thick wedge of atmosphere, making the sunlight much less intense. The high albedo, because of ice and snow, reflects a good portion of the sun's light.

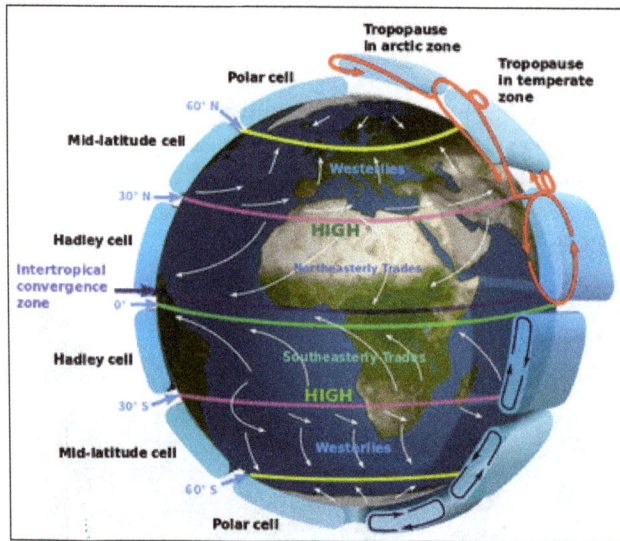

Atmospheric Circulation

The position of a region relative to the circulation cells and wind belts has a great effect on its climate. In an area where the air is mostly rising or sinking, there is not much wind.

Intertropical Convergence Zone

The Intertropical Convergence Zone (ITCZ) is the low pressure area near the equator in the boundary between the two Hadley Cells. The air rises so that it cools and condenses to create clouds and rain. Climate along the ITCZ is therefore warm and wet. Early mariners called this region the doldrums because their ships were often unable to sail because there were no steady winds.

The ITCZ migrates slightly with the season. Land areas heat more quickly than the oceans. Because there are more land areas in the Northern Hemisphere, the ITCZ is influenced by the heating effect of the land. In Northern Hemisphere summer, it is approximately 5° north of the equator while in the winter it shifts back and is approximately at the equator. As the ITCZ shifts, the major wind belts also shift slightly north in summer and south in winter, which causes the wet and dry seasons in this area.

Hadley Cell and Ferrell Cell Boundaries

At about 30 degrees N and 30 degrees S, the air is fairly warm and dry because much of it came from the equator where it lost most of its moisture at the ITCZ. At this location the air is descending, and sinking air warms and causes evaporationMariners named this region the horse latitudes. Sailing ships were sometimes delayed for so long by the lack of wind that they would run out of water and food for their livestock. Sailors tossed horses and other animals over the side after they died. Sailors sometimes didn't make it either.

Prevailing Winds

The prevailing winds are the bases of the Hadley, Ferrell, and Polar Cells. These winds greatly influence the climate of a region because they bring the weather from the locations they come from. For example, in California, the predominant winds are the westerlies blowing in from the Pacific Ocean, which bring in relatively cool air in summer and relatively warm air in winter. Local winds also influence local climate. For example, land breezes and sea breezes moderate coastal temperatures.

Continental Position

When a particular location is near an ocean or large lake, the body of water plays an extremely important role in affecting the region's climate:

- A maritime climate is strongly influenced by the nearby sea. Temperatures vary a relatively small amount seasonally and daily. For a location to have a true maritime climate, the winds must most frequently come off the sea.

- A continental climate is more extreme, with greater temperature differences between day and night and between summer and winter.

The ocean's influence in moderating climate can be seen in the following temperature comparisons. Each of these cities is located at 37 °N latitude, within the westerly winds.

Maritime climate.

Continental climate.

Ocean Currents

The temperature of the water offshore influences the temperature of a coastal location, particularly if the winds come off the sea. The cool waters of along the western United States is caused by a clockwise rotating ocean current that is bringing cold water from the arctic toward the equator. The climatic effect is that coastal regions of California, Oregon, and Washington are are cool. Coastal upwelling also brings cold, deep water up to the ocean surface off of California, which contributes to the cool coastal temperatures. But that same ocean current brings warm, tropical water to eastern Japan. In the Atlantic Ocean, the northern ocean current, called the Gulf Stream, brings warm water from the tropics to the southern states. This is a major reason why the southern states experience humid conditions in the summer and tornadoes because of all this warm moisture. The Gulf Stream also impacts Europe by bringing warm water northward, making this region that is rather northward warmer than expected.

Altitude and Mountain Ranges

Air pressure and air temperature decreases with altitude. The closer molecules are packed together, the more likely they are to collide. Collisions between molecules give off heat, which warms the air. At higher altitudes, the air is less dense and air molecules are more spread out and less likely to collide. A location in the mountains has lower average temperatures than one at the base of the mountains. In Colorado, for example, Lakewood (5,640 feet) average annual temperature is 62 degrees F (17 degrees C), while Climax Lake (11,300 feet) is 42 degrees F (5.4 degrees C). Mountain ranges have two effects on the climate of the surrounding region. The first is something called the rainshadow effect, which brings warm dry climate to the leeward size of a mountain range. The second effect mountains have on climate systems is the ability to separate coastal regions from the rest of the continent. Since a maritime air mass may have trouble rising over a mountain range, the coastal area will have a maritime climate but the inland area on the leeward side will have a continental climate.

Climate Zones and Biomes

A climate zone results from the climate conditions of an area: its temperature, humidity, amount and type of precipitation, and the season. A climate zone is reflected in a region's natural vegetation. Perceptive travelers can figure out which climate zone they are in by looking at the vegetation, even if the weather is unusual for the climate on that day.

The major factors that influence climate determine the different climate zones. In general, the same type of climate zone will be found at similar latitudes and in similar positions on nearly all continents, both in the Northern and Southern Hemispheres. The one exception to this pattern is the climate zones called the continental climates, which are not found at higher latitudes in the Southern Hemisphere. This is because the Southern Hemisphere land masses are not wide enough to produce a continental climate.

The most common system used to classify climatic zones is the Köppen classification system. This system is based on the temperature, the amount of precipitation, and the times of year when precipitation occurs. Since climate determines the type of vegetation that grows in an area, vegetation is used as an indicator of climate type.

A climate type and its plants and animals make up a biome. The organisms of a biome share certain characteristics around the world, because their environment has similar advantages and challenges. The organisms have adapted to that environment in similar ways over time. For example, different species of cactus live on different continents, but they have adapted to the harsh desert in similar ways.

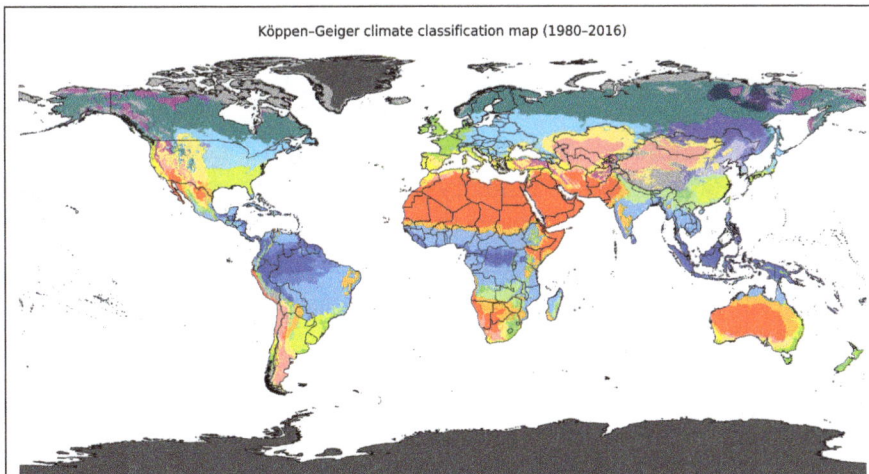

Köppen–Geiger climate classification map (1980–2016)

The Köppen classification system recognizes five major climate groups, each with a distinct capital letter A through E. Each lettered group is divided into subcategories. Some of these subcategories are forest (f), monsoon (m), and wet/dry (w) types, based on the amount of precipitation and season when that precipitation occurs.

Tropical Moist Climates (Group A)

Tropical Moist (Group A) climates are found in a band about 15° to 25° N and S of the equator. What climate characteristics is the tropical moist climate group likely to have?

- Temperature: Intense sunshine; each month has an average temperature of at least 18 °C (64 °F).

- Rainfall: Abundant, at least 150 cm (59 inches) per year.

The wet tropics have almost no annual temperature variation and tremendous amounts of rain fall year round, between 175 and 250 cm (65 and 100 inches). These conditions support the tropical rainforest biome. Tropical rainforests are dominated by densely packed, broadleaf evergreen trees. These rainforests have the highest number of species or biodiversity of any ecosystem.

Tropical Monsoon (Am)

The tropical monsoon climate has very low precipitation for one to two months each year. Rainforests grow here because the dry period is short, and the trees survive off of soil moisture. This climate is found where the monsoon winds blow, primarily in southern Asia, western Africa, and northeastern South America.

Tropical Wet and Dry

The tropical wet and dry climate lies between about 50 and 200 latitude, around the location of the ITCZ. In the summer, when the ITCZ drifts northward, the zone is wet. In the winter, when the ITCZ moves toward the equator, the region is dry. This climate exists where strong monsoon winds blow from land to sea.

Rainforests cannot survive the months of low rainfall, so the typical vegetation is savanna. This biome consists mostly of grasses, with widely scattered deciduous trees and rare areas of denser forests.

Dry Climates (Group B)

The Dry Climates (Group B) have less precipitation than evaporation. Dry climate zones cover about 26% of the world's land area. What climate characteristics is the dry climate group likely to have?

- Temperature: Abundant sunshine. Summer temperatures are high; winters are cooler and longer than Tropical Moist climates.

- Rainfall: Irregular; several years of drought are often followed by a single year of abundant rainfall.

Arid Desert (Bw)

Low-latitude, arid deserts are found between 15° to 30 °N and S latitudes. This is where warm dry air sinks at high pressure zones. True deserts make up around 12% of the world's lands.

In the Sonoran Desert of the southwestern United States and northern Mexico, skies are clear. The typical weather is extremely hot summer days and cold winter nights. Although annual rainfall is less than 25 cm (10 inches), rain falls during two seasons. Pacific storms bring winter rains and monsoons bring summer rains. Since organisms do not have to go too many months without some rain, a unique group of plants and animals can survive in the Sonoran desert.

Semi-arid or Steppe (Bs)

Higher latitude semi-arid deserts, also called steppe, are found in continental interiors or in rainshadows. Semi-arid deserts receive between 20 and 40 cm (8 to 16 inches) of rain annually. The annual temperature range is large. In the United States, the Great Plains, portions of the southern California coast and the Great Basin are semi-arid deserts.

Moist Subtropical Mid-latitude (Group C)

What climate characteristics is the moist subtropical group likely to have?

- Temperature: The coldest month ranges from just below freezing to almost balmy, between -3 °C and 18 °C (27° to 64 °F). Summers are mild with average temperatures above 10 °C (50 °F). Seasons are distinct.

- Rainfall: There is plentiful annual rainfall.

Dry Summer Subtropical or Mediterranean Climates (cs)

The Dry Summer Subtropical climate is found on the western sides of continents between 30° and 45° latitude. Annual rainfall is 30 to 90 cm (14 to 35 inches), most of which comes in the winter.

The climate is typical of coastal California, which sits beneath a summertime high pressure for about five months each year. Land and sea breezes make winters moderate and summers cool. Vegetation must survive long summer droughts. The scrubby, woody vegetation that thrives in this climate is called chaparral.

Humid Subtropical (Cfa)

The Humid Subtropical climate zone is found mostly on the eastern sides of continents. Rain falls throughout the year with annual averages between 80 and 165 cm (31 and 65 inches). Summer days are humid and hot, from the lower 30's up to 40 °C (mid-80's up to 104 °F). Afternoon and evening thunderstorms are common. These conditions are caused by warm tropical air passing over the hot continent. Winters are mild, but middle-latitude storms called cyclones may bring snow and rain. The southeastern United States, with its hot humid summers and mild, but frosty winters, is typical of this climate zone.

Marine West Coast Climate (Cfb)

This climate lines western North America between 40° and 65° latitude, an area known as the Pacific Northwest. Ocean winds bring mild winters and cool summers. The temperature range, both daily and annually, is fairly small. Rain falls year round, although summers are drier as the jet stream moves northward. Low clouds, fog, and drizzle are typical. In Western Europe the climate covers a larger region since no high mountains are near the coast to block wind blowing off the Atlantic.

Humid Continental (Group D)

Continental (Group D) climates are found in most of the North American interior from about 40 °N to 70 °N. What climate characteristics is the continental group most likely to have?

- Temperature: The average temperature of the warmest month is higher than 10 °C (50 °F) and the coldest month is below -3 °C (-27 °F).

- Precipitation: Winters are cold and stormy. Snowfall is common and snow stays on the ground for long periods of time.

Trees grow in continental climates, even though winters are extremely cold, because the average annual temperature is fairly mild. Continental climates are not found in the Southern Hemisphere because of the absence of a continent large enough to generate this effect.

Humid Continental (Dfa, Dfb)

The humid continental climates are found around the polar front in North America and Europe. In the winter, middle-latitude cyclones bring chilly temperatures and snow. In the summer, westerly winds bring continental weather and warm temperatures. The average July temperature is often above 20 °C (70 °F). The region is typified by deciduous trees, which protect themselves in winter by losing their leaves.

The two variations of this climate are based on summer temperatures:

- Dfa, long, hot summers: Summer days may be over 38 °C (100 °F), nights are warm and the temperature range is large, perhaps as great as 31 °C (56 °F). The long summers and high humidity foster plant growth.

- Dfb, long, cool summers: Summer temperatures and humidity are lower. Winter temperatures are below -18 °C (0 °F) for long periods.

Subpolar (Dfc)

The subpolar climate is dominated by the continental polar air that masses over the frigid continent. Snowfall is light, but cold temperatures keep snow on the ground for months. Most of the approximately 50 cm (20 inches) of annual precipitation falls during summer cyclonic storms. The angle of the Sun's rays is low but the Sun is visible in the sky for most or all of the day during the summer, so temperatures may get warm, but are rarely hot. These continental regions have extreme annual temperature ranges. The boreal, coniferous forests found in the subpolar climate are called taiga and have small, hardy, and widely spaced trees. Taiga vast forests stretch across Eurasia and North America.

Polar Climates (Group E)

Polar climates are found across the continents that border the Arctic Ocean, Greenland, and Antarctica. What climate characteristics is the polar climate group most likely to have?

- Temperature: Winters are entirely dark and bitterly cold. Summer days are long, but the sun is low on the horizon so summers are cool. The average temperature of the warmest month at less than 10 °C (50 °F). The annual temperature range is large.

- Precipitation: The region is dry with less than 25 cm (10 inches) of precipitation annually; most precipitation occurs during the summer.

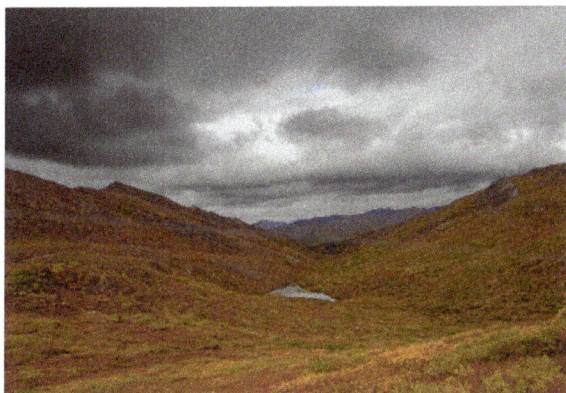

Polar Tundra (ET)

The polar tundra climate is continental, with severe winters. Temperatures are so cold that a layer of permanently frozen ground, called permafrost forms below the surface. This frozen layer can extend hundreds of meters deep. The average temperature of the warmest months is above freezing, so summer temperatures defrost the upper-most portion of the permafrost. In winter, the permafrost prevents water from draining downward. In summer, the ground is swampy. Although the precipitation is low enough in many places to qualify as a desert, evaporation rates are also low, so the landscape receives more usable water than a desert.

Because of the lack of ice-free land near the South Pole, there is very little tundra in the Southern Hemisphere. The only plants that can survive the harsh winters and soggy summers are small ground-hugging plants like mosses, lichens, small shrubs, and scattered small trees that make up the tundra.

Ice Cap

Ice caps are found mostly on Greenland and Antarctica, about 9% of the Earth's land area. Ice caps may be thousands of meters thick. Ice cap areas have extremely low average annual temperatures, e.g. -29 °C (-20 °F) at Eismitte, Greenland. Precipitation is low because the air is too cold to hold much moisture. Snow occasionally falls in the summer.

Hightland Climates (Group H)

When climate conditions in a small area are different from those of the surroundings, the climate of the small area is called a microclimate. The microclimate of a valley may be cool relative to its surroundings since cold air sinks. The ground surface may be hotter or colder than the air a few feet above it, because rock and soil gain and lose heat readily. Different sides of a mountain will have different microclimates. In the Northern Hemisphere, a south-facing slope receives more solar energy than a north-facing slope, so each side supports different amounts and types of vegetation.

Altitude mimics latitude in climate zones. Climates and biomes typical of higher latitudes may be found in other areas of the world at high altitudes.

Climate has changed throughout Earth history. Much of the time Earth's climate was hotter and more humid than it is today, but climate has also been colder, as when glaciers covered much more of the planet. The most recent ice ages were in the Pleistocene Epoch, between 1.8 million and 10,000 years ago. Glaciers advanced and retreated in cycles, known as glacial and interglacial periods. With so much of the world's water bound into the ice, sea level was about 125 meters (395 feet) lower than it is today. Many scientists think that we are now in a warm, interglacial period that has lasted about 10,000 years.

For the past 2,000 years, climate has been relatively mild and stable when compared with much of Earth's history. Why has climate stability been beneficial for human civilization? Stability has allowed the expansion of agriculture and the development of towns and cities.

Fairly small temperature changes can have major effects on global climate. The average global temperature during glacial periods was only about 5.5 degrees C (10 degrees F) less than Earth's current average temperature. Temperatures during the interglacial periods were about 1.1 degrees C (2.0 degrees F) higher than today.

Since the end of the Pleistocene, the global average temperature has risen about 4 degrees C (7 degrees F). Glaciers are retreating and sea level is rising. While climate is getting steadily warmer, there have been a few more extreme warm and cool times in the last 10,000 years. Changes in climate have had effects on human civilization.

- The Medieval Warm Period from 900 to 1300 A.D. allowed Vikings to colonize Greenland and Great Britain to grow wine grapes.

- The Little Ice Age, from the 14th to 19th centuries, the Vikings were forced out of Greenland and humans had to plant crops further south.

Short-term Climate Changes

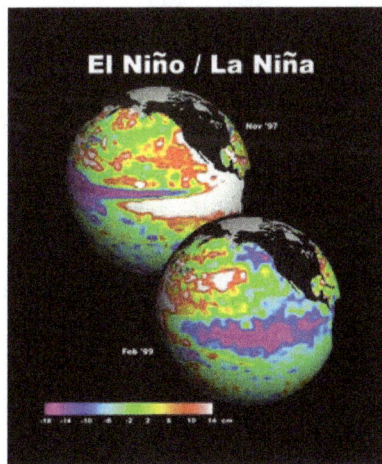

Short-term changes in climate are common. The largest and most important of these is the oscillation between El Niño and La Niña conditions. This cycle is called the ENSO

(El Niño southern oscillation). The ENSO drives changes in climate that are felt around the world about every two to seven years.

In a normal year, the trade winds blow across the Pacific Ocean near the equator from east to west (toward Asia). A low pressure cell rises above the western equatorial Pacific. Warm water in the western Pacific Ocean and raises sea levels by one-half meter. Along the western coast of South America, the Peru Current carries cold water northward, and then westward along the equator with the trade winds. Upwelling brings cold, nutrient-rich waters from the deep sea.

In an El Niño year, when water temperature reaches around 28 degrees C (82 degrees F), the trade winds weaken or reverse direction and blow east (toward South America). Warm water is dragged back across the Pacific Ocean and piles up off the west coast of South America. With warm, low-density water at the surface, upwelling stops. Without upwelling, nutrients are scarce and plankton populations decline. Since plankton form the base of the food web, fish cannot find food, and fish numbers decrease as well. All the animals that eat fish, including birds and humans, are affected by the decline in fish.

By altering atmospheric and oceanic circulation, El Niño events change global climate patterns:

- Some regions receive more than average rainfall, including the west coast of North and South America, the southern United States, and Western Europe.

- Drought occurs in other parts of South America, the western Pacific, southern and northern Africa, and southern Europe.

An El Niño cycle lasts one to two years. Often normal circulation patterns resume. Sometimes circulation patterns bounce back quickly and extremely, called La Niña.

In a La Niña year, as in a normal year, trade winds moves from east to west and warm water piles up in the western Pacific Ocean. Ocean temperatures along coastal South America are colder than normal (instead of warmer, as in El Niño). Cold water reaches farther into the western Pacific than normal.

Other important oscillations are smaller and have a local, rather than global, effect. The North Atlantic Oscillation mostly alters climate in Europe. The Mediterranean also goes through cycles, varying between being dry at some times, and warm and wet at others.

Causes of Long-term Climate Change

Many processes can cause climate to change. These include changes in the amount of energy the Sun produces over years; the positions of the continents over millions of years; in the tilt of Earth's axis; orbit over thousands of years; that are sudden and dramatic because of random catastrophic events, such as a large asteroid impact; in greenhouse gases in the atmosphere, caused naturally or by human activities.

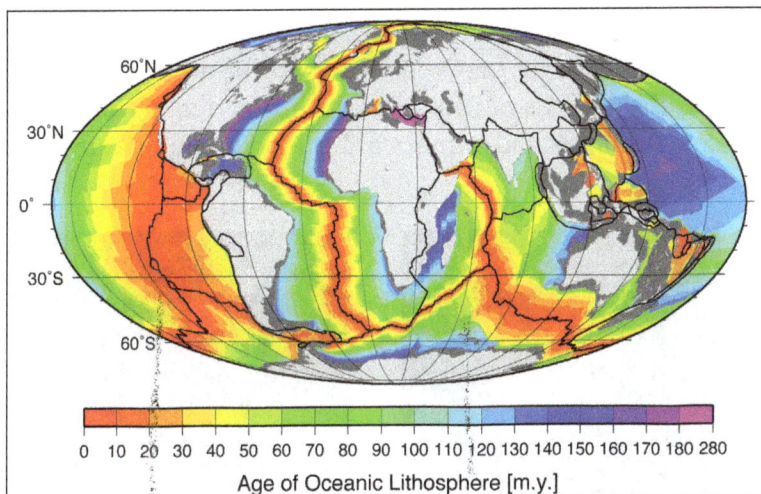

Age of Oceanic Lithosphere [m.y.]

Plate Tectonics

Plate tectonic movements can alter climate. Over millions of years as seas open and close, ocean currents may distribute heat differently. For example, when all the continents are joined into one supercontinent (such as Pangaea), nearly all locations experience a continental climate. When the continents separate, heat is more evenly distributed. Plate tectonic movements may help start an ice age. When continents are located near the poles, ice can accumulate, which may increase albedo and lower global temperature. Low enough temperatures may start a global ice age.

Plate motions trigger volcanic eruptions, which release dust and CO_2 into the atmosphere. Ordinary eruptions, even large ones, have only a short-term effect on weather. Massive eruptions of the fluid lavas that create lava plateaus release much more gas and dust, and can change climate for many years. This type of eruption is exceedingly rare; none has occurred since humans have lived on Earth.

Milankovitch Cycles

The most extreme climate of recent Earth history was the Pleistocene. Scientists attribute a series of ice ages to variation in the Earth's position relative to the Sun, known as Milankovitch cycles. The Earth goes through regular variations in its position relative to the Sun. The shape of the Earth's orbit changes slightly as it goes around the Sun. The orbit varies from more circular to more elliptical in a cycle lasting between 90,000 and 100,000 years. When the orbit is more elliptical, there is a greater difference in solar radiation between winter and summer. The planet wobbles on its axis of rotation. At one extreme of this 27,000 year cycle, the Northern Hemisphere points toward the Sun when the Earth is closest to the Sun. Summers are much warmer and winters are much colder than now. At the opposite extreme, the Northern Hemisphere points toward the Sun when it is farthest from the Sun. This results in chilly summers and warmer winters.

The planet's tilt on its axis varies between 22.1 degrees and 24.5 degrees. Seasons are caused by the tilt of Earth's axis of rotation, which is at a 23.50 angle now. When the tilt angle is smaller, summers and winters differ less in temperature. This cycle lasts 41,000 years.

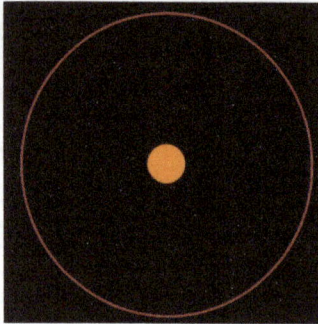

Eccentricity – circular.   Eccentricity – elliptical.

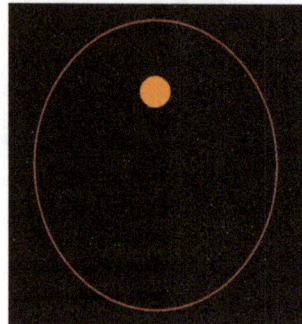

When these three variations are charted out, a climate pattern of about 100,000 years emerges. Ice ages correspond closely with Milankovitch cycles. Since glaciers can form only over land, ice ages only occur when landmasses cover the polar regions. Therefore, Milankovitch cycles are also connected to plate tectonics.

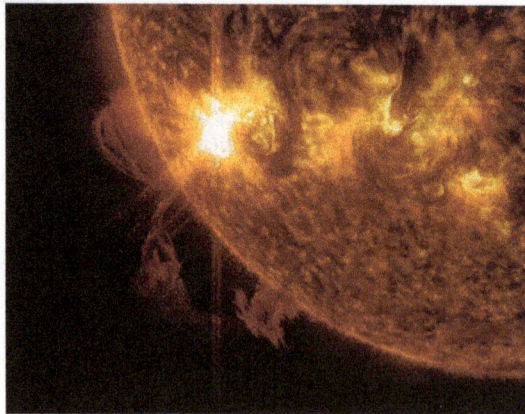

Sun Variation

The amount of energy the Sun radiates is variable. Sunspots are magnetic storms on the Sun's surface that increase and decrease over an 11-year cycle. When the number of sunspots is high, solar radiation is also relatively high. But the entire variation in solar radiation is tiny relative to the total amount of solar radiation that there is and there is no known 11-year cycle in climate variability. The Little Ice Age corresponded to a time when there were no sunspots on the Sun.

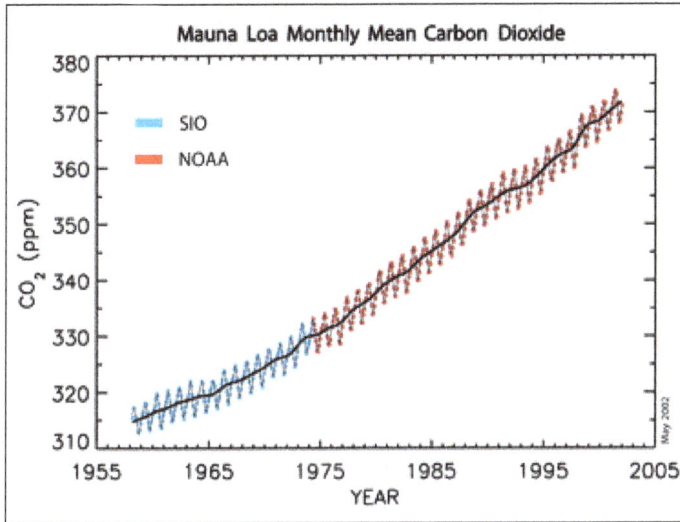

Mauna Loa Monthly Mean Carbon Dioxide

Changes in Atmospheric Greenhouse Gas Levels

Since greenhouse gases trap the heat that radiates off the planet's surfaces what would happen to global temperatures if atmospheric greenhouse gas levels decreased? What if greenhouse gases increased? A decrease in greenhouse gas levels decreases global temperature and an increase raises air temperature. Greenhouse gas levels have varied throughout Earth history. For example, CO_2 has been present at concentrations less than 200 parts per million (ppm) and more than 5,000 ppm. But for at least 650,000 years, CO_2 has never risen above 300 ppm, during either glacial or interglacial periods. Natural processes add (volcanic eruptions and the decay or burning of organic matter) and remove absorption by plants, animal tissue, and the ocean) CO_2 from the atmosphere. When plants are turned into fossil fuels the CO_2 in their tissue is stored with them. So CO_2 is removed from the atmosphere. What does this do to Earth's average temperature?

Fossil fuel use has skyrocketed in the past few decades more people want more cars and industrial products. This has released CO_2 into the atmosphere. Burning tropical rainforests, to clear land for agriculture, a practice called slash-and-burn agriculture, also increases atmospheric CO_2. By cutting down trees, they can no longer remove CO_2 from the atmosphere. Burning the trees releases all the CO_2 stored in the trees into the atmosphere.

There is now nearly 40% more CO_2 in the atmosphere than there was 200 years ago, before the Industrial Revolution. About 65% of that increase has occurred since the first CO_2 measurements were made on Mauna Loa Volcano, Hawaii, in 1958. CO_2 is the most important greenhouse gas that human activities affect because it is so abundant. But other greenhouse gases are increasing as well. A few are:

- Methane: Released from raising livestock, rice production, and the incomplete burning of rainforest plants.

- Chlorofluorocarbons (CFCs): Human-made chemicals that were invented and used widely in the 20th century.

- Tropospheric ozone: From vehicle exhaust, it has more than doubled since 1976.

Anthropogenic Climate Change

The United States has long been the largest emitter of greenhouse gases, with about 20% of total emissions in 2004. As a result of China's rapid economic growth, its emissions surpassed those of the United States in 2008. However, it's also important to keep in mind that the United States has only about one-fifth the population of China. What's the significance of this? The average United States citizen produces far more greenhouse gases than the average Chinese person.

If nothing is done to decrease the rate of CO_2 emissions, by 2030, CO_2 emissions are projected to be 63% greater than they were in 2002.

If nothing is done to control greenhouse gas emissions and they continue to increase at current rates, the surface temperature of the Earth can be expected to increase between 0.5 degrees C and 2.0 degrees C (0.9 degrees F and 3.6 degrees F) by 2050 and between 2 degrees and 4.5 degrees C (3.5 degrees and 8 degrees F) by 2100, with CO_2 levels over 800 parts per million (ppm). On the other hand, if severe limits on CO_2 emissions begin soon, temperatures could rise less than 1.1 degrees C (2 degrees F) by 2100.

Glaciers are melting and vegetation zones are moving uphill. If fossil fuel use exploded in the 1950s, why do these changes begin early in the animation? Does this mean that the climate change we are seeing is caused by natural processes and not by fossil fuel use?

As greenhouse gases increase, changes will be more extreme. Oceans will become slightly more acidic, making it more difficult for creatures with carbonate shells to grow, and that includes coral reefs. A study monitoring ocean acidity in the Pacific Northwest found ocean acidity increasing ten times faster than expected and 10% to 20% of shellfish (mussels) being replaced by acid tolerant algae.

Plant and animal species seeking cooler temperatures will need to move poleward 100 to 150 km (60 to 90 miles) or upward 150 m (500 feet) for each 1.0 degrees C (8 degrees F) rise in global temperature. There will be a tremendous loss of biodiversity because forest species can't migrate that rapidly. Biologists have already documented the extinction of high-altitude species that have nowhere higher to go.

Decreased snow packs, shrinking glaciers, and the earlier arrival of spring will all lessen the amount of water available in some regions of the world, including the western United States and much of Asia. Ice will continue to melt and sea level is predicted to rise 18 to 97 cm (7 to 38 inches) by 2100. An increase this large will gradually flood coastal regions where about one-third of the world's population lives, forcing billions of people to move inland.

Although scientists do not all agree, hurricanes are likely to become more severe and possibly more frequent. Tropical and subtropical insects will expand their ranges, resulting in the spread of tropical diseases such as malaria, encephalitis, yellow fever, and dengue fever.

You may notice that the numerical predictions above contain wide ranges. Sea level, for example, is expected to rise somewhere between 18 and 97 cm — quite a wide range. What is the reason for this uncertainty? It is partly because scientists cannot predict exactly how the Earth will respond to increased levels of greenhouses gases. How quickly greenhouse gases continue to build up in the atmosphere depends in part on the choices we make.

An important question people ask is this: Are the increases in global temperature natural? In other words, can natural variations in temperature account for the increase in temperature that we see? The scientific data shows no, natural variations cannot explain the dramatic increase in global temperatures. Changes in the Sun's irradiance, El Niño and La Niña cycles, natural changes in greenhouse gas, plate tectonics, and the Milankovitch Cycles cannot account for the increase in temperature that has already happened in the past decades.

In December 2013 and April 2014, the Intergovernmental Panel on Climate Change (IPCC) released a series of damaging reports on not only the current scientific knowledge of climate change, but also on the vulnerability and impacts to humans and ecosystems. But it is important to get a strong, data driven understanding of climate change. Along with the IPCC, other organizations like the United Nations Environmental Programme (UNEP), World Health Organizations, World Meteorological Organization (WMO), NASA, the National Oceanic and Atmospheric Administration (NOAA), and the U.S. Environmental Protection Agency (EPA).

For the past two centuries, climate has been relatively stable. People placed their farms and cities in locations that were in a favorable climate without thinking that the climate could change. But climate has changed throughout Earth history, and a stable climate is not the norm. In recent years, Earth's climate has begun to change again. Most of this change is warming because of human activities that release greenhouse gases into the atmosphere. The effects of warming are already being seen and will become more extreme as temperature rise.

Temperatures are Rising

With more greenhouse gases trapping heat, average annual global temperatures are rising. This is known as global warming. While temperatures have risen since the end of the Pleistocene, 10,000 years ago, this rate of increase has been more rapid in the past century, and has risen even faster since 1990. The nine warmest years on record have all occurred since 1998, and the 10 of the 11 warmest years have occurred since 2001 (through 2012). The 2000s were the warmest decade yet. Annual variations aside, the average global temperature increased about 0.8 degrees C (1.5 degrees F) between 1880 and 2010, according to the Goddard Institute for Space Studies, NOAA.

Future Warming

The amount CO_2 levels will rise in the next decades is unknown. What will this number depend on in the developed nations? What will it depend on in the developing nations? In the developed nations it will depend on technological advances or lifestyle changes that decrease emissions. In the developing nations, it will depend on how much their lifestyles improve and how these improvements are made. Computer models are used to predict the effects of greenhouse gas increases on climate for the planet as a whole and also for specific regions.

Whatever the temperature increase, it will not be uniform around the globe. A rise of 2.8 degrees C (5 degrees F) would result in 0.6 degrees to 1.2 degrees C (1 degree to 2 degrees F) at the equator, but up to 6.7 degrees C (12 degrees F) at the poles. So far, global warming has affected the North Pole more than the South Pole, but temperatures are still increasing at Antarctica.

The timing of events for species is changing. Mating and migrations take place earlier in the spring months. Species that can are moving their ranges uphill. Some regions that were already marginal for agriculture are no longer farmable because they have become too warm or dry.

Weather will become more extreme with heat waves and droughts. Some modelers predict that the Midwestern United States will become too dry to support agriculture and that Canada will become the new breadbasket. In all, about 10% to 50% of current cropland worldwide may become unusable if CO_2 doubles. There are global monitoring systems to help monitor potential droughts that could turn into famines if they occur in politically and socially unstable regions of the world and if appropriate action isn't taken in time. One example is FEWS NET, which is a network of social and environmental scientists using geospatial technology to monitor these situations. But even with proper monitoring, if nations don't act, catastrophes can occur like in Somalia from 2010-2012.

Components of Climate Change

Simple Model of Global Radiation Balance

The earth receives energy from the sun in the form of visible, near-infrared, and ul-traviolet radiation. Most of this energy is either reflected back to space by clouds and other bright surfaces (about 30%), or is absorbed by the earth's surface. A significant fraction of the near-infrared component is absorbed by water vapor. However, since the greatest amount of vapor is found at low levels near the surface, that is where the near-infrared is preferentially absorbed in the atmosphere. Ultraviolet radiation, which forms only a small fraction of the energy coming from the sun, is mostly absorbed by ozone in the stratosphere.

Figure below illustrates a simple model of the gross radiation balance of the earth. The solar flux at the earth's orbit is F_s =1370 W m^{-2}. If a fraction A = 0.3 is reflected by the earth, then the amount of solar radiation absorbed by the earth is equal to $F_s(1 - A)$ times the projected area of the earth πR^2 , where R = 6370 km is the earth's radius. This amounts to 122 PW. (One petawatt = 1 PW = 10^{15} W.)

This input of energy is balanced by the outow of thermal radiation. The radiative temperature of the earth T_{rad} is the temperature which would result in this outow if the earth radiated like a black body:

$$\pi R^2 F_s (1 - A) = 4\pi R^2 \sigma T_{rad}^4$$

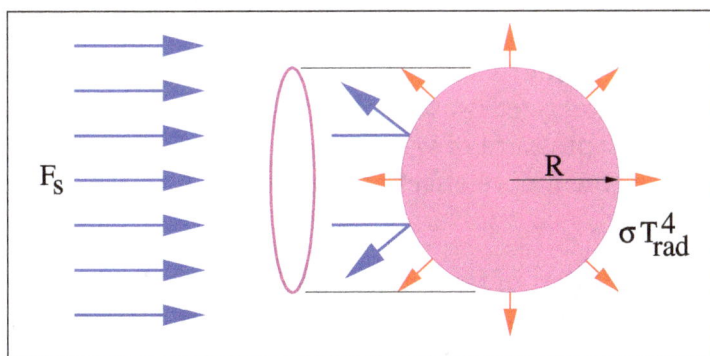

Gross radiation balance of the earth. Incoming solar radiation with flux F_s must be balanced by losses from reflection and infrared black body emission.

Where $\sigma = 5.67 \times 10^{-8} W\ m^{-2} K^{-4}$. Solving For T_{rad} Results in,

$$T_{rad} = \left[\frac{F_s(1-A)}{4\sigma} \right]^{\frac{1}{4}}$$

which for the given numbers yields T_{rad} = 255 K.

The radiative temperature thus calculated is significantly colder than the mean temperature of the earth's surface, which is near 288 K. This is because the earth's atmosphere is somewhat opaque to the thermal infrared radiation emitted by the surface of the earth, which means that the actual radiation to space occurs not from the surface, but from some higher level in the atmosphere, which is colder. The opacity of the atmosphere to thermal infrared radiation has little to do with the main atmospheric constituents, nitrogen and oxygen, but is caused primarily by the trace gases carbon dioxide, water vapor, methane, and especially in the stratosphere, ozone. In the troposphere, water vapor is the most important of these gases. The elevation of the surface temperature above the radiative temperature is called the greenhouse effect.

Stability of Earth's Climate

Equilibrium does not necessarily imply stability. Just because the earth is near a state of balance between incoming and outgoing energy doesn't mean that it will remain near this state. We now explore two simple "toy" models of energyflow which exhibit instability. The idea here is not to assert that the earth behaves in the manner indicated by the model, but only to indicate the breadth of possibilities.

Runaway Greenhouse Effect

The difference between the surface temperature of the earth and its radiative temperature is due to the blocking of the radiative loss to space from the earth's surface by greenhouse gases in the atmosphere. Water vapor plays a large role in this blocking effect. However, as the surface temperature increases, the amount of water vapor in the atmosphere is likely to increase. It is reasonable to assume that the relative humidity of the atmosphere would remain unchanged as the surface gets warmer. the column-integrated water vapor per unit area would be proportional to the saturation vapor pressure of water vapor at the surface temperature. Under these conditions, the greenhouse effect of water vapor would increase with the surface temperature.

The ocean covers a large part of the earth's surface and it stores most of the thermal energy in the climate system. An approximate equation for the time rate of change of mean ocean temperature T_{ocean} (assuming that the ocean covers the entire globe) is,

$$\rho_w D C_1 \frac{dT_{ocean}}{dt} \equiv F_n = \frac{F_s(1-A)}{4} - \sigma \left(T_{ocean} - \delta T_G \right)^4$$

where F_n is the net flux imbalance, ρ_w is the density of sea water, D is the mean depth of the ocean (or the ocean layer being considered), C_l is the specific heat of sea water and $\delta T_G = T_{ocean} - T_{rad}$ is the greenhouse gain, i. e., the difference between the ocean temperature and the earth's radiative temperature. In the above steady state model we found that $\delta T_G \equiv \delta T_{GR} = 33$ K.

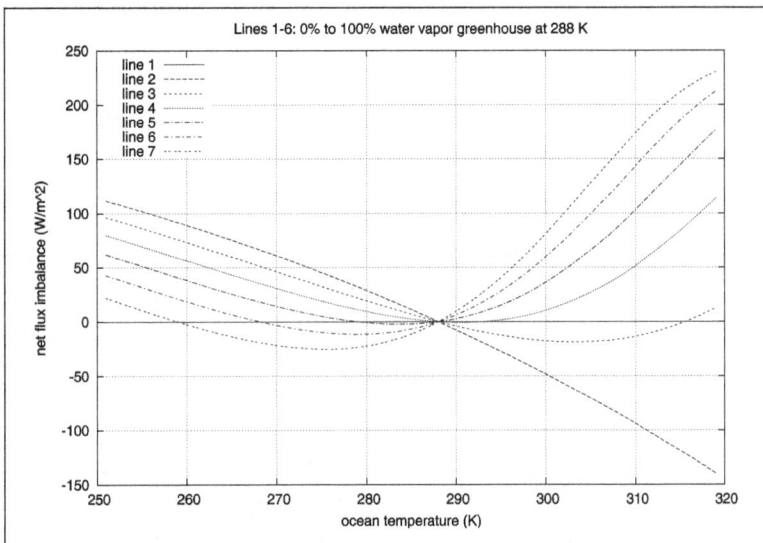

Net Flux imbalance as a function of ocean temperature for fixed albedo A = 0.3.
Lines 2 – 7 represent values of the water vapor fraction f = 0.0, 0.2, 0.4, 0.6, 0.8, and 1.0.

Let us assume that f is the fraction of the greenhouse gain which is caused by water vapor at the reference ocean temperature T_R, which we take to be 288 K. This part of the greenhouse gain we set proportional to the saturation vapor pressure at the temperature of the ocean, while the remaining part we assume to be independent of temperature. An approximate formula for the saturation vapor pressure of water vapor is,

$$e_s(T) \approx e_s(T_R) \exp\left[\frac{L}{R_v} \left(\frac{1}{T_R} - \frac{1}{T_{ocean}} \right) \right]$$

Where T_R is a constant reference temperature, L is the latent heat of condensation for water, and $R_v = R_{univ}/m_w$, where R_{univ} is the universal gas constant and m_w is the molecular weight of water vapor. We therefore set:

$$\delta T_G = \left(1 - f\right) \delta T_{GR} + f \delta T_{GR} \exp\left[\frac{L}{R_v} \left(\frac{1}{T_R} - \frac{1}{T_{ocean}} \right) \right]$$

Which yields $\delta T_G = \delta T_{GR}$ when $T_{ocean} = T_R$. Inserting this into equation yields an expression for the net flux imbalance F_n.

Figure shows a plot of the net flux imbalance as a function of ocean temperature for fixed albedo A = 0.3 and variable water vapor greenhouse fraction f. Intersection of a curve with the F_n = 0 line indicates an equilibrium point. The stability of this equilibrium is given by the slope of the line passing through this point-a negative slope indicates that a higher than equilibrium temperature results in cooling while a temperature lower than equilibrium results in heating. In this situation the ocean temperature relaxes back toward the equilibrium point, corresponding to stable equilibrium. On the other hand, a positive slope means that a temperature slightly warmer than equilibrium causes heating, etc., which corresponds to unstable equilibrium. Unstable equilibrium results in a runaway effect, either toward further heating or cooling, depending whether the initial temperature perturbation from equilibrium is positive or negative.

In this model the current equilibrium energy balance of the earth is unstable if the water vapor greenhouse fraction f > 0.4. In such a situation a slight increase in ocean temperature results in further increases without bound as the column-integrated water vapor increases and the atmosphere becomes more opaque to infrared radiation. This is the runaway greenhouse effect.

Clearly the current climate system is not in a state of unstable equilibrium, since any small perturbation to the system would have led long ago to either a runaway greenhouse or to a return to the lower temperature stable equilibrium point. According to this theory, the current climate can only be stable if the water vapor greenhouse fraction f < 0.4. This constitutes a bit of a mystery, as we believe that water vapor has a larger role than this in the greenhouse effect. It turns out that this model is significantly over simplified - we shall return to this issue later.

Ice World

So far we have assumed that the albedo of the earth is constant. However, the albedo can change as the amount of clouds and the amount of snow or ice cover changes. One hypothesis is that the ice age earth is a stable state because the albedo of an ice-covered earth should be much higher than that of the earth as it is today. A high albedo would have the effect of reducing the absorbed incoming solar radiation, resulting in a lower radiative temperature for the earth. We will now insert a simple temperature-dependent formula for albedo into equation (1.3) with a fixed greenhouse gain δT_G = 33 K:

$$A = 0.45 - 0.3 \tan^{-1}\left[\left(T_{ocean} - 270\right)/5\right]/\pi$$

The albedo asymptotes to 0.6 for $T_{ocean} \ll 270$ K, and to 0.3 for $T_{ocean} \ll 270$ K, with A = 0.45 at T_{ocean} = 270 K, which is near the freezing point of the ocean.

Figure below shows how the net flux imbalance varies with ocean temperature. Equilibrium states exist at three temperatures, 259 K, 273 K, and 284 K. However, only the First and the third of these are stable equilibrium points - the second is unstable, as inspection of figure shows. This system is therefore bistable, and with sufficiently strong pushes in the right direction, can be induced to switch from one state to the other. Our current state is near the upper equilibrium point. We call the lower equilibrium point "ice world", since the earth would be covered by glaciers under these conditions. This state may have something to do with the ice ages.

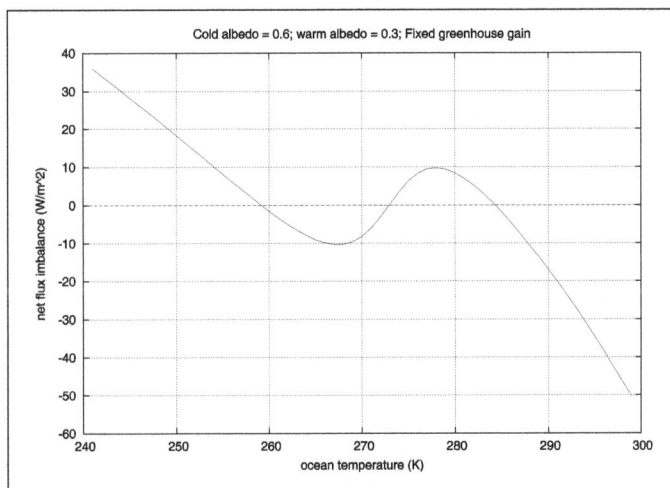

Net flux imbalance for the case of variable albedo, equal to 0.6 for $T_{ocean} \ll 270$ K and to 0.3 for $T_{ocean} \gg 270$ K.

Latitudinal Energy Transfer

Energy deposition is not uniform with latitude. Less solar radiation is deposited at high latitudes than low, and in the winter than in the summer hemisphere. A local equilibrium temperature can be computed at each latitude, but the resulting temperature distribution has a much steeper decline toward the poles than is observed. Thus, energy must be transported from the tropical regions toward the poles.

Let us make a quantitative calculation of this effect for the case in which the sun is directly over the equator, i. e., at the equinox. The key issue is the actual versus the projected area of a latitudinal strip of the earth's surface, as illustrated in figure. The actual area of the strip of earth's surface illustrated in this figure is $\delta S = 2\pi R \cos\phi \cdot R\delta\phi$, while the projected area of this strip as seen from the sun is $\delta S_p = 2 R \cos\phi \cdot R \cos\phi \, \delta\phi$. Assuming albedo A at latitude ϕ, the energy balance at this latitude is $Fs(1-A)\delta S_p = \sigma T_{rad}^4 \delta S$, resulting in a radiative temperature there of,

$$T_{rad}(\phi) = \left[\frac{F_s(1-A)\cos\phi}{\pi\sigma} \right]^{1/4}$$

Figure below shows the latitudinal distribution of radiative temperature as well as the global radiative temperature and the mean sea surface temperature as a function of latitude. Also plotted is the sea surface temperature minus 40 K, slightly greater than the difference between the surface temperature and the radiative temperature in the globally uniform case. This can be taken as an approximation of the actual local radiative temperature. Within 50° of the equator the predicted radiative temperature exceeds the actual value, where as, at higher latitudes the reverse is true. This implies lateral export of energy by the oceans and atmosphere from low latitudes to high, i. e., transport of energy down the temperature gradient. In other words, there is a net flow of energy into the atmosphere and oceans at low latitudes, followed by transport to high latitudes, where there is net export.

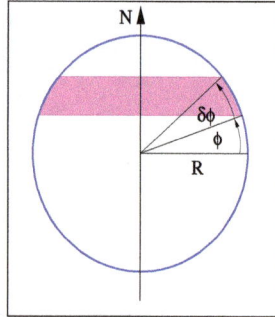

Sketch of geometry used in the calculation of the
latitudinal dependence of radiative temperature at the equinox.

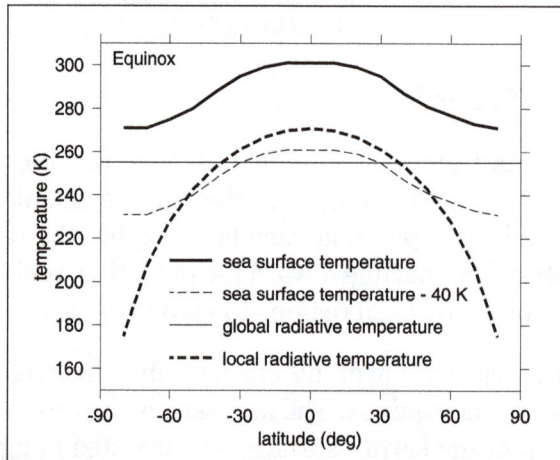

Annual and longitudinal mean of sea surface temperature, sea surface temperature −40 K,
global mean radiative temperature and latitudinal distribution of radiative temperature at the equinox.

Figure above shows that this transport actually does take place. On an annual average, import of energy by radiation at the top of the atmosphere exceeds export by 2 + 5 + 2.5 = 9.5 PW between 30 °S and 30 °N. The same amount is exported to space at higher latitudes. The poleward transport of energy is shared almost equally by the atmosphere and the ocean, with the ocean contributing slightly more.

Within 10° of the equator, the atmospheric absorption of solar radiation and the emission of infrared are nearly in balance, so that the net absorption is only about 2 PW. The absorbed energy is exported to higher latitudes. This compares with a solar input of

$F_s(1-A) \cdot 2R^2 \cos^2 \varphi \delta\varphi$, which equals 27 PW for $\varphi = 0$ and $\delta\varphi = 20/57.3$ radians. Thus, the atmosphere in this band exports laterally only about 7% of the incoming solar radiation. However, an additional 11%, or 3 PW of incoming solar energy travels indirectly to higher latitudes via the oceans.

The transition between net inflow from space to net outflow to space occurs near latitudes ±30°. This is lower in latitude than suggested by the estimate in figure where this transition occurs nearer 50°. However, we must remember that figure is based on rather loose arguments.

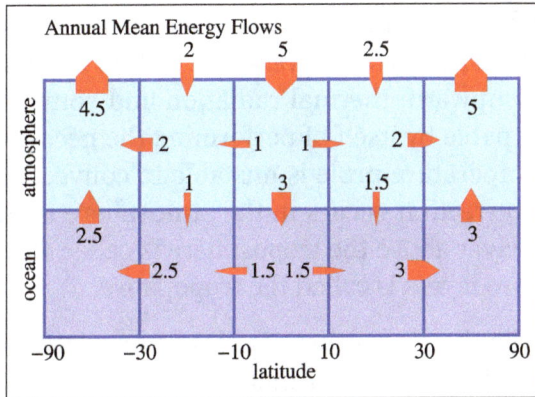

Annually averaged net energy flows in petawatts (10^{15} W; fluid transport plus radiation) between space and various latitudinal segments of the atmosphere and ocean.

Figure below shows the global energy flows as in figure , except averaged over December, January, and February only, i. e., during the northern winter. As would be expected from the southerly position of the sun during this period, there is a net inflow of energy into the southern hemisphere and a net outflow in the northern hemisphere. These hemispheric imbalances are partially compensated by flow of energy from south to north in both the atmosphere and the ocean. However, this flow doesn't account for all of the southern hemisphere gains and northern hemisphere losses. Substantial warming with time occurs in the southern hemisphere as well as substantial cooling north of the equator. This heat storage effect is most important in the oceans, as the oceanic heat capacity is much higher than that of the atmosphere. The situation for the northern summer is nearly a mirror image of that for the northern winter.

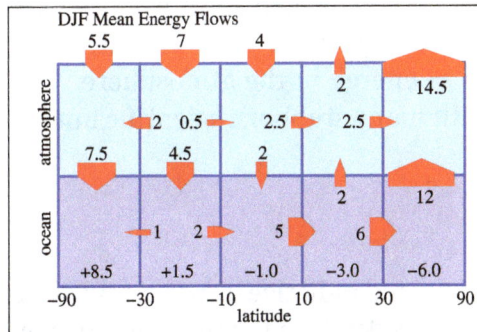

December-January-February averaged net energy flows (in petawatts; fluid transport plus radiation) between space and various latitudinal segments of the atmosphere and ocean. The numbers at the bottom are the rates at which energy is stored in the ocean segments. Zero storage is assumed for the atmosphere.

Vertical Energy Transfer

Atmosphere

Solar energy is deposited at the surface and at low levels in the atmosphere, while departing thermal infrared radiation is emitted on the average at atmospheric temperatures of 255 K, which corresponds to an average elevation of about 6 km. Two processes transport this energy upward, thermal radiation and convection. It turns out that radiative transfer is incapable by itself of performing the necessary transport, because the resulting vertical temperature prole is unstable to convection below a certain level. The layer in which convection occurs in the atmosphere is called the troposphere. The convectively stable layer above the troposphere is called the stratosphere, and the boundary between the two layers is called the tropopause.

In the stratospheric layer the atmosphere is close to radiative equilibrium, i. e., the vertical radiative energy fluxes are such that zero net heating occurs at each level in this layer. The tropical atmosphere within 10° of the equator has small lateral fluxes compared to the upward and downward energy fluxes due to solar radiation, thermal radiation, and convection. Thus, the predominant balance in the tropical troposphere is nearly one of radiative-convective equilibrium, i. e., one in which convective heating at each level is balanced by radiative cooling.

Since the surface of the earth is nearly covered by water, the earth's atmosphere contains a great deal of water vapor. Since rising convective parcels cool in their ascent, they eventually reach a level, called the lifting condensation level (LCL), at which the water vapor in the parcel begins to condense. This has three consequences. First, the condensation releases latent heat, resulting in higher temperatures in ascent than in a parcel without condensation. Second, much of this condensed water falls out of the parcel as precipitation. As a result, the parcel follows a different thermodynamic trajectory in the descending part of its convective cycle. Third, the small water drops and ice crystals which form when a parcel reaches the LCL and the freezing level interact strongly with solar and thermal radiation, thus modifying the radiative transfer of energy in the atmosphere. The net result is a complex and interesting system with many stubborn scientific uncertainties of importance to weather and climate.

Ocean

Water is highly opaque to infrared radiative transfer, and even in the clearest ocean water most of the solar radiation is deposited in the upper 100 m or so. Convection occurs in the ocean in regions where the high density of cold, salty surface water is sufficient to produce negative parcel buoyancies. This sinking water becomes the source of deep ocean water. The upward return circulation occurs at low latitudes far from the sinking motion, which occurs only in preferred regions at high latitudes. The complete circulation system is called the thermohaline circulation. Our knowledge of the details of this circulation is quite sketchy at this point. In particular, the mechanisms governing the upwelling of deep water are not well understood.

Just as the budget of moisture is inextricably intertwined with the budget of energy in the atmosphere, the budget of salt plays a crucial role in the energy budget in the ocean. This is because salt content has a large effect on ocean water density. Salt content is increased as a result of evaporation at the ocean surface, and is decreased by the inow of fresh water into the ocean from rivers and precipitation.

The salt content of the ocean is measured in terms of salinity S. Salinity is defined officially in terms of the electrical conductivity of ocean water. However, in practical terms the salinity is just the mass of dissolved salts per unit mass of water, measured in grams of salt per kilogram of fresh water.

Hydrological Cycle

The Intergovernmental Panel on Climate Change (IPCC) warned, in 2001, that the global climate is changing, largely because of human activities. The recognition that climate change could pose a serious threat to the sustainability of our current way of life has led to increasing interest in climate system research. Hydrotechnical engineers can contribute greatly to this growing field of study, since they are well-suited to examine the strong connections between the climate system and the hydrological cycle. However, engineers first require a greater understanding of the feedbacks between these two systems. This understanding is crucial, because changes in the climate will trigger long-term, and potentially extensive, changes in the hydrological cycle, with significant impacts on society and on the environment.

Following the approach of the IPCC, considers the climate system as composed of five separate components: atmosphere, oceans, cryosphere, land surface, and biosphere. Each component receives a fairly thorough discussion, which focuses on its climatically important characteristics and processes, and the discussion of the atmospheric component explains the basis of anthropogenic climate change. After each component is described individually, a discussion of feedbacks explains the importance of both connections between climate system components, and spatial and temporal scale in determining system behaviour. Two examples of important feedbacks in the climate system are given – both form integral parts of the global hydrological cycle.

In addition to explaining the basis of general circulation model (GCM) use with hydrological applications, an alternative approach is provided, which promotes the use of simple but comprehensive, rather than complex but climate-only models. Such an approach cannot provide high resolution results for hydrological studies, but it includes the social, economic, and environmental context of climate change, which is beyond the capabilities of GCMs.

Climate System Components and Feedbacks

Our understanding of the climate system is based on long term observations and modelling work. Climate monitoring provides an observational record of the past and current state of the climate, including mean state and variability. Such information is necessary if we are to understand climate change and its causes, track the results of climate change adaptation and mitigation efforts, and improve and validate climate models. Modelling serves a very different purpose. It allows us to test and improve our understanding of the physical processes that drive the climate system, to identify the feedbacks that operate within and between climate system components, and to predict future changes in regional and global climate.

In climate system research, the atmosphere and the oceans have generally received the most attention from the climate modelling community, which has focused primarily on simulating the processes within and between these two parts of the overall system. While initial work on general circulation models (GCM) considered the atmosphere and oceans separately, resulting in the creation of individual Atmospheric GCMs (AGCMs) and Ocean GCMs (OGCMs), the focus has shifted over the past decade or so to coupling the two through the mass, heat, and momentum exchanges, or flows, between them. In such coupled models, there is also a growing recognition that the cryosphere, the land surface, and the biosphere play crucial roles in the climate system through feedbacks to the atmosphere and the oceans.

Atmosphere

One of the two fluid components of the climate system, the atmosphere is characterized by organized circulation, chaotic motions, and random turbulence, and forms the most variable and rapidly changing part of the climate system. The atmosphere has characteristics that vary with height, from sea level to an altitude of roughly 100 km. These changes in physical properties result in a heightbased division of the atmosphere into four parts, two of which are relevant to climate change: the troposphere, (from sea level to 10-15 km in altitude), and the stratosphere (from the top of the troposphere, to about 50 km altitude).

The troposphere contains the majority of the earth's weather, and is fundamentally driven by surface heating, which results in "the convective overturning of air [that] characterizes the region". The atmospheric circulation in the troposphere depends on imbalances between radiative heating at low- and high-latitudes. This uneven heating leads to a distribution of mass that "drives a meridional overturning of air, with rising motion at low latitudes and sinking motion at high latitudes". The atmosphere also circulates latitudinally, because of the Earth's rotation, so that at middle and high latitudes, the general thermal structure approximately parallels the latitudinal circles. The majority of the longitudinal (meridional) heat distribution is based on asymmetries in the instantaneous circulation, which leads to heat exchange between the tropics and the poles. At low latitudes, planetary rotation plays a smaller role; instead, the geographical distribution of atmospheric heating determines atmospheric circulation via latent heating and the resulting Walker (east-west overturning) and Hadley (north-south overturning) circulations. As a consequence of the atmospheric general circulation, temperatures at fixed altitudes in the troposphere decrease poleward from the equator, where they are at a maximum. Furthermore, pressure is only 10% of its surface value at an altitude of 10 km, while temperature falls almost linearly at a lapse rate of 6 K km-1 to roughly 220 K (-53 °C) at the same altitude. The stratosphere differs significantly from the troposphere because of weak vertical motions and strong radiative processes. Although generally neglected in the past, the IPCC reports that stratospheric effects "can have a detectable and perhaps significant influence on tropospheric climate".

The climate varies naturally, but the current changes in the climate system result from anthropogenic emissions of greenhouse gases to the atmosphere. As a whole, the atmosphere consists primarily of nitrogen and oxygen (78% and 21% by volume, respectively), while argon (0.93%), water vapour, carbon dioxide, ozone, and several other gases make up the remaining 1% of the atmospheric volume. Nitrogen, oxygen, and argon are not greenhouse gases; however, trace gases like carbon dioxide (CO_2), methane, nitrous oxide, and others are, and therefore both absorb and emit infrared radiation, making them an important part of the Earth's energy balance. Furthermore, water vapour and ozone, although not directly emitted like the other trace gases, are also highly important, natural greenhouse gases with concentrations that vary according to atmospheric feedbacks. These greenhouse gases allow solar radiation to pass through the atmosphere freely but capture the infrared radiation emitted by the Earth, thereby performing a critical service to all life on Earth – without them the planet would be roughly 33°C cooler, with an average temperature of -18 °C rather than a comfortable 15 °C.

So the problem is not the greenhouse effect, per se. Instead, the current change in climate results from an enhanced greenhouse effect, caused by an increase in the atmospheric concentration of anthropogenic trace gases. Increases in greenhouse gases, and particularly in CO_2, enhance absorption and emission of infrared radiation and lead to an increase in the opacity of the atmosphere. Since the Earth emits longwave radiation essentially like a blackbody radiator, and an increase in opacity shifts the height at which radiation is emitted to space toward higher, colder altitudes, this decrease in emitting-temperature leads to a reduction in the outgoing longwave radiation, and a consequent increase in global surface temperature. Overall, the imbalance between incoming and outgoing radiation, called a radiative forcing, is very small, at 4 W m-2 of forcing for a doubling of CO_2. Without any further feedbacks, this forcing would result in a surface-troposphere temperature increase of roughly 1.2 °C. However, the increase in global surface temperatures triggers a series of feedbacks within and between different climate system components, which result in a temperature increase of between 1.5 °C and 4.5 °C. These feedbacks relate the atmosphere to all of the other climate components.

Oceans

As the second of the two fluid components of the climate system, the oceans are more stable than the atmosphere. Oceanic circulation depends on both winds and density contrasts caused by thermal and salinity gradients, and occurs more slowly than the atmospheric circulation. Basically, water in tropical regions stores a great deal of heat from the warm, tropical atmosphere, which ocean currents then transport towards the poles. At higher latitudes, the transported water cools, warming the cooler atmosphere there – this kind of circulation reduces the pole-to-equator temperature gradient. The strongest currents generally take the form of Western Boundary Currents, such as the Gulf Stream, Kuroshio, Agulhas Current, and others, which transport much of the heat and fresh water in the climate system. They are mainly wind-driven, with typical widths of 50 km.

The oceans consist of two vertical layers, the surface mixed layer, and the ocean interior. The mixed layer is generally more buoyant than water in the ocean interior, being warmer and less salty, and a thermocline, lying at 100 m to 1000 m depth, normally separates the layers. Subduction and convection form the basis of exchanges between the surface and interior. Subductive processes work as follows. In sufficient quantities, turbulent energy can increase the depth of the mixed layer through entrainment of deeper water. However, upwelling at the base of the mixed layer, from Ekman pumping, can cause the deepened mixed layer to cool quickly, leading to a shallowing of the mixed layer. When this occurs, water that passes the deepest point of the mixed-layer is transferred to the ocean interior and is subducted. Larger horizontal gradients in the mixed layer depth lead to larger subduction rates, meaning that "variations in mixed layer depth are of primary importance in setting the structure of the interior of the ocean".

Convection plays a large role in determining the overall circulation of the oceans, and generally depends on seasonal differences in water temperature. At high latitudes during the winter, surface water loses its buoyancy and becomes denser than the water below. The result is mixing of the surface layer with the ocean interior at highly variable mixing depths, which reach their maximum at the end of the cooling season. During the summer, the ocean surface warms, creating a shallow mixed layer that isolates the newly formed deep water from the atmosphere, which currents and mesoscale eddies (at a scale of 50 to 100 km) then transfer into the abyssal ocean. Overall, deep convective mixing "constitutes a very efficient vertical transfer process". For example, the Labrador Sea, a major site of open ocean deep convection, replenishes the deep waters of the Atlantic, Pacific, and Indian Oceans. Related to deep convective mixing, shelf convection occurs near continents, and is likely the source of the densest bottom water, Antarctic Bottom Water (AABW), which circulates through all three major oceans. Deep water heating, via diapycnal mixing across density surfaces in the ocean interior, allows the cold water to rise through the thermocline, slowly returning it to the ocean surface.

Cryosphere

The cryosphere consists of all the snow and ice in the climate system, and "includes the ice sheets of Greenland and Antarctica, continental glaciers and snow fields, sea ice, and permafrost". It has several characteristics that are important to the climate system, including its high reflectivity of solar radiation (albedo), low thermal conductivity, and high thermal inertia, and its storage of large quantities of fresh water.

On the land, snow cover changes the Earth's albedo and surface roughness, and influences water transfer between snow and soil. Permafrost may affect the climate by releasing greenhouse gases as it thaws, and vegetation changes may influence the radiation balance and surface hydrology as the active layer above permafrost thickens. Land ice also takes the form of glaciers and ice sheets, both of which are formed by the burial and densification of snow. Glaciers come in many shapes and sizes; when they become "continental scale masses of fresh water ice", they are called ice sheets, which exist either entirely on land or have areas that are afloat.

Sea ice is impossible to describe without including climate system feedbacks. It serves several purposes in the climate system, moderating the heat exchange between the ocean and atmosphere, transporting fresh water in the ocean, and modifying the surface albedo. Sea ice at higher latitudes influences the extent of visible ocean surface and affects the albedo. Furthermore, ice thickness and open leads (gaps) determine heat releases to the atmosphere and subsequent temperature increases. Finally, sea ice formation can alter oceanic deep water production through the process of brine rejection – brine affects surface water density. Because denser water sinks, changes to this process could have "an influence on the water mass structure that stretches far beyond the area of sea ice".

Land Surface

According to Baede et al., land surface characteristics of relevance to the global climate include vegetation cover and soil properties, land surface texture, and the amount of dust, which affects atmospheric radiative processes when wind-borne. Vegetation cover and soil qualities influence the transfer of solar energy from the ground to the atmosphere, as infrared radiation or as latent (evapotranspiration) heat – soil moisture content influences surface temperatures strongly through the energy used in evaporating water. Land surface texture, or roughness, affects atmospheric dynamics through its influence on winds; mountains can also have large effects on the climate at regional and continental scales, or even global scales.

Climate modellers and hydrological engineers view the land surface fundamentally differently. This difference in focus has a simple root: hydrological engineers interact more directly with society than do climate modellers, working to mitigate water related hazards (floods, droughts, and landslides), and to improve "agriculture and food production, human health, municipal and industrial supply, and environmental quality". They also tend to work on hydrological issues at much smaller spatial – catchment or municipality, rather than ocean, continent or globe – and temporal – hours, days, years, rather than seasons, decades, centuries – scales, and require much more location-specific data than modellers do.

For climate modellers, the important processes and characteristics of the land surface are those that act at large scales to influence the global climate, and particularly the atmosphere and oceans, through latent and sensible heat transfers in the atmosphere; snow, ice, or vegetation-cover albedo effects; and water transfers from the continents into the cryosphere or the oceans. Some land-surface schemes now include river routing, improving runoff-modelling from large (10^5 km^2 or greater) drainage basins. However, the climate modelling community has begun to improve its representation of land surface processes, realizing that accurate modelling of atmosphere-land surface feedbacks is critical for realistic simulation of continental climate and hydrology.

Hydrological engineers are interested in precipitation, evaporation, water storage and runoff, which affect soil moisture, groundwater resources, river flows, and lake levels, and are affected in turn by soil structure, soil chemistry, vegetation, climate and topography. All of these hydrological processes and characteristics play a role in water resources management issues such as "irrigation and drainage, hydropower, flood control, water supply, and inland navigation". On a large scale, "precipitation drives the continental hydrology, and influences the salinity of the ocean". It is, however, highly variable in both space and time: stratiform precipitation, with a spatial resolution of up to and beyond 100 km, dominates in the extra-tropics except over the continents in summer, whereas convective precipitation, with a horizontal scale of only a few kilometres, dominates in the tropics. Evaporation occurs from open water, shallow groundwater, and water stored on vegetation, and depends on available energy, the moisture content of the air, and air movement; as surface temperatures increase, the level of potential evaporation tends to rise as well, because warmer air holds more water. Water storage in soil influences the rate of actual evaporation and the rate of runoff, and depends on soil properties. Finally, runoff occurs when precipitation is greater than the sum of the evapotranspiration and water storage in a catchment. Interestingly, with the exception of water storage in aquifers and ice, all of these important hydrological processes involve feedbacks from one component of the climate system to another.

Biosphere

The biosphere includes both terrestrial and oceanic life. Terrestrial life plays an important role in the global carbon cycle, the hydrological cycle, and the surface albedo, all of which involve feedbacks between different climate system components. Oceanic life helps to moderate atmospheric CO_2 concentrations. In terms of the carbon cycle, apparent enhancements of photosynthetic versus respiration rates in the last several decades imply that the global carbon cycle could strongly influence the future concentrations of atmospheric carbon dioxide, and thus reduce both the rate and extent of global warming. However, temperature increases associated with higher atmospheric CO_2 concentrations may also enhance microbial decomposition of soil matter, resulting in releases of CO_2 – views are mixed on this subject.

On time scales of millennia, the oceans determine atmospheric CO_2 concentrations through two fundamental processes: the solubility pump and the biological pump. Oceanic convection transfers surface water containing dissolved CO_2 into the deep ocean, providing an effective, long-term means of carbon storage in the ocean interior; however, CO_2 is most soluble in cold, saline water, so the solubility pump may become less effective as surface temperatures rise. The biological pump may provide an alternative. Phytoplankton use carbon dioxide for photosynthesis, which lowers the partial pressure of CO_2 in the upper ocean and promotes its absorption from the atmosphere. Furthermore, certain species of phytoplankton and zooplankton form dense calcium carbonate shells. When they die, their shells 'rain' down into the deeper water – this carbonate pump therefore also results in CO_2 transfer to the ocean interior.

In terms of the hydrological cycle, increasing atmospheric CO_2 concentrations allow vegetation to maintain the same photosynthetic rate at a lower evapotranspiration rate, which may result in a decrease in transpiration. However, CO_2-fertilization causes plants to grow faster and larger, and may counteract CO_2-driven increases in vegetation's water use efficiency. Unfortunately, the effects of increasing carbon dioxide on terrestrial vegetation are therefore difficult to predict.

Finally, terrestrial vegetation can influence local albedo strongly, potentially affecting not only regional, but even global climate. For example, many hope that increased forest cover at high latitudes will help to sequester CO_2, thereby reducing its atmospheric concentration; however, since forests are generally darker than open land, the resulting lower surface albedo may cause boreal and cool-temperate forests to warm the climate. Overall, vegetation effects on climate are simply not straightforward, since boreal forests tend to warm the environment through decreased albedo, while tropical forests generally cool and moisten the local climate through evapotranspiration.

Feedback Processes

Issues of connectivity and scale are critical in feedback relationships. Clearly, although climate system feedbacks exist both within particular system components and between them, the descriptions of climate system components above focussed, where possible, on the processes and characteristics relevant to individual parts of the overall system. In explaining the components separately, the aim was twofold: first, to improve understanding of the climatically-important elements of each system component; and second, to demonstrate how climate behaviour arises from, and depends on, links between components. The Earth system functions as a whole, not as separate parts. Feedbacks between components are critical to understanding climatic behaviour, because the system is complex. According to Rind, "a complex system is literally one in which there are multiple interactions between many different components" – a description that matches the climate system perfectly. However, despite its complexity, the climate is not entirely unpredictable, and typically exhibits linear behaviour to small changes in external forcings. In the highly improbable circumstances that climate forcings exceed certain thresholds, the climate system undergoes dramatic reorganizations. Overall, then, connectivity issues form the basis of understanding and predicting the response of the climate system to anthropogenic forcings.

Feedbacks can connect small to large scales. However, spatial and temporal scales are generally important for a different reason. They determine which feedbacks to include in climate models, and which to neglect. Essentially, the feedbacks included in a model depend on the model's purpose. For example, in modelling global-mean temperature, the accuracy of regional precipitation predictions is unimportant, since GCMs that cannot predict regional climate already simulate surface air temperature "particularly well with nearly all models closely matching the observed magnitude of variance and exhibiting a correlation >0.95 with the observations"; however, if the flooding of a small catchment is of interest, local rainfall predictions are clearly critical. This topic therefore describes feedbacks that are important at global versus regional scales – a discussion more useful to hydrological engineers than one dealing with differences in temporal feedback scales between months and millennia. It turns out that two of the most important feedbacks at any scale relate directly to the hydrological cycle: the water vapour feedback, and the cloud feedback.

Water vapour is the strongest greenhouse gas: its associated feedback approximately doubles the climate warming from what it would be with fixed atmospheric levels. At present, the amount of water vapour in the atmosphere can certainly increase, as the majority of the troposphere is highly under saturated with respect to water. The feedback functions as follows. Water vapour enters the atmosphere through evaporation from either the land surface or from the oceans, and the lateral transport of water through the atmosphere supplies the necessary moisture for precipitation and runoff. Once in the atmosphere, increases in the "abundance and vertical distribution of water vapour [cause very strong interactions] with convection and cloudiness, thereby influencing the albedo of the planet as well as the infrared opacity of the atmosphere". In general, large spatial scales apply to water vapour feedbacks, because of the large-scale atmospheric circulations that influence the horizontal water vapour concentrations – for example, "less than 20% of the precipitation that falls comes from evaporation within a distance of about 1000 km". However, along with exchanges of energy, moisture transfers between soil, vegetation, snow, and the overlying atmosphere determine a large part of regional climate. Overall, the water vapour feedback, as just one part of the hydrological cycle, involves interaction between the atmosphere, oceans, land surface, and biosphere.

Clouds and water vapour are directly connected. Clouds serve as both sources and sinks for water vapour, while water vapour provides the foundation for cloud formation. Like water vapour, clouds have strong effects on the climate system; unlike water vapour, the cloud response to climate change "remains a dominant source of uncertainty" in climate modelling and prediction. Indeed, at present there is not even a theoretical basis for predicting the sign of cloud-cover feedback. Overall, the difficulty with the cloud feedback is its complexity: while the large scale atmospheric processes determine the formation and evolution of clouds, short temporal and spatial scales control important radiative and latent heating processes that feed back to the global scale. For example, buoyancy, moisture and condensation interact on scales of millimetres to tens of kilometres to determine the behaviour of critical convective processes. Latitude and altitude also play a role: at low latitudes, clouds warm the surface by increasing infrared radiation absorption and reducing longwave emissions to space, but also cool the surface by directly reflecting solar radiation to space; at high latitudes, they seem to operate in the opposite fashion. To confuse the issue further, high, cold clouds warm the atmosphere, while low clouds cool the atmosphere, especially at high latitudes. And yet, in spite of this complexity, it is important that the cloud feedback is understood, because cloud feedbacks "are likely to control the bulk precipitation efficiency and associated responses of the planet's hydrological cycle to climate radiative forcings". Cloud process modelling at various scales has provided considerable insight into the cloud feedback.

Models of the Climate System

At least three main categories of climate models exist, with their classification based on both model resolution and purpose. Simple models have low temporal and spatial resolution, and produce zonally- or globally-averaged results for temperature, but not for other variables such as rainfall; however, they are computationally fast and their behaviour is easy to understand. Earth System Models of Intermediate Complexity (EMIC) sit in the middle of the climate model range, being simpler and lower resolution than GCMs, and therefore running much faster. However, they are also much more comprehensive than simple models, and explicitly couple components of the Earth system. Finally, General Circulation Models, or Global Climate Models (GCM), calculate "the full threedimensional character of the climate comprising at least the global atmosphere and the oceans. The solution of a series of equations that describe the movement of energy, momentum, and various tracers and the conservation of mass is therefore required". Of the three model types, GCMs provide the most realistic and complete representation of climate.

Current coupled GCMs join atmospheric and oceanic components with generally simple sea-ice and landsurface process models. These models run, with typical time-steps of 30 minutes, at variable resolutions on a three-dimensional grid: AGCMs typically employ horizontal resolutions of roughly 250 km and 10 to 30 vertical levels, while typical OGCMs have horizontal resolutions of 125 to 250 km, and vertical resolutions of 200 to 400 m. Overall, GCMs produce credible simulations of annual mean climate and seasonal cycles that are useful even at the sub-continental scale, but their resolution is insufficient to provide regional predictions of climate change, mostly because of poor simulation of land-surface processes. They therefore cannot simulate regional rainfall, much less extreme events: "Extreme precipitation is difficult to reproduce, especially for the intensities and patterns of heavy rainfall which are heavily affected by the local scale". Indeed, with grid resolutions generally greater than 200 km, models cannot even resolve many precipitation-related processes, which must therefore be parameterized.

GCM results are extremely hard to evaluate because of model complexity, which places severe limits on analyzing and understanding the models' processes, interactions, and uncertainties. Simulations require so much computational power and time that only limited numbers of multidecadal experiments can be performed, and model intercomparison projects have trouble obtaining the necessary broad participation. However, such intercomparisons are crucial, because there is no 'best' model, making combinations of results from many coupled models necessary for accurate prediction of climate change effects.

Linking Climate and Hydrological Models

Although climate change will certainly affect the hydrological cycle, its results will not be distributed uniformly over the globe. Because of the potentially drastic consequences for societies and ecosystems of water scarcity or overabundance, an increasing number of researchers have attempted to determine the results of climate change in specific regions, or for certain river basins. In general, they have used at least four approaches to assessing climate change impacts on catchments and river basins, including "estimates obtained by applying arbitrary changes in climate input to hydrological models, spatial analogue techniques, temporal analogue techniques, and the use of results from GCMs, either directly or by downscaling to the appropriate catchment scale".

Of the four options, GCM-derived results present the best choice for hydrological model inputs, although the application of a GCM to regional or smaller scales presents certain difficulties, the most obvious of which lies in the incompatibilities between model scales. The variables provided by GCM simulations rarely match those required by hydrological applications – GCMs can provide, for example, mean river flow, mean and seasonal groundwater recharge, and mean seasonal (monthly) variation in river flow, but they cannot directly give flow-duration curves of river flow, run-sums for reservoirs, mean annual flood, and so on. Unfortunately, even the direct use of GCM data is problematic, as it often gives "inaccurate hydrological simulations due to the different temporal and spatial scales used in GCMs and hydrological models".

Manabe et al., Wood et al., and Arora and Boer describe applications of GCMs to hydrological modelling. Manabe et al. attempt to determine the effects of quadrupled atmospheric CO_2 concentrations on river discharge and soil moisture, using a relatively coarse-resolution coupled GCM and a hydrological 'bucket' model, in which a bucket represents the moisture-holding capacity of each continental grid box. They find that water-rich areas will receive more water, and water-poor areas will lose more water under climate change, and discuss their results in terms of climatic processes and feedbacks, rather than comparing them with observations. Wood et al. compare forecasts of streamflow and other seasonal hydrological variables obtained from climate models versus forecasts formed via resampling of climate observations, and obtain variable results: "while climate model forecasts presently suffer from a general lack of skill, there may be locations, times of year, and conditions (e.g. during El Nino or La Nina) for which they improve hydrological forecasts". Arora and Boer (2001) analyze the hydrology of 23 major river basins as atmospheric CO_2 concentrations increase by 1% year-1 from the present until 2100. They determine the effects of climate change on basin-wide averaged mean annual precipitation, evapotranspiration, and discharge, monthly streamflow at the river mouths, and changes in extreme flow events; however, because of differences in mean annual precipitation between model simulations and observations, "the values of mean annual discharge from the control simulation do not always compare well with observations" – note that the simulated mean annual runoff was within 20% of the observed estimates for only 4 of the 23 river basins.

Downscaling provides an alternative to the problems posed by using GCMs to drive hydrological models directly. Although a variety of techniques exist, including statistical-empirical, dynamical, and statisticaldynamical methods, describes dynamical downscaling, because of its close connection to GCMs and climate modelling. Dynamical downscaling involves the use of a fine-resolution model, called a Regional Circulation Model (RCM), to provide driving-fields of climatic variables to hydrological models. Most RCMs are applied at a spatial resolution of 30 to 100 km to a grid-box side; however, some research groups have run multi-year simulations at a resolution of between 10 and 20 km. Since most RCMs simulate regions of the Earth, rather than the whole globe, they require input at their lateral boundaries from either GCMs or observational data. They certainly offer advantages over coarser models in predicting regional climate, such as temperature biases of only about 20 °C, and precipitation biases of about 50% of observations, for models with 50 km horizontal resolutions that are driven by climate observations. However, RCMs also have drawbacks, including high computational cost, sensitivity to errors in the driving GCM, and a lack of feedback from RCM to GCM, since RCMs are usually 'one-way nested' in GCMs. Furthermore, when driving-fields from RCM climate simulations form the input to hydrological models, the same feedback issues arise, especially if GCMs drive RCMs, and RCMs drive hydrological models in turn, with no feedbacks from any finer- to coarser-resolution models.

Overall, RCMs provide better forcings for hydrological models than GCMs do. Kim investigates the effects of climate change on extreme events in the Western United States, using a 36-km resolution RCM forced by a single GCM simulation. He describes the changes in hydrological variables associated with an increase in atmospheric CO_2, and finds overall that climate change will increase extreme events in the region, both in terms of the number of wet days and the mean intensity of the events, and that the changes will occur most strongly in mountainous regions. Kleinn et al. simulate streamflows in the Rhine basin, using a set of two RCMs that were driven by weather observations, coupled to a fineresolution (1 km grid-size) hydrological model. Their modelling configuration includes a coarse resolution RCM (56 km grid-size) and its nested fine-resolution RCM (14 km grid-size), "a distributed hydrological runoff model, and a downscaling interface for the one-way coupling of the two models". They find that, although the finer model captures regional precipitation patterns better than its coarser counterpart, there are no appreciable differences in the simulated hydrological outputs, probably because of the aggregation of precipitation in the hydrological model. Finally, Andreasson et al. simulate the results of climate change on six river-basins in Sweden, using two GCMs to provide climatic driving-fields for two RCMs, which drive a hydrological model in turn. They explain that changes in large-scale circulation patterns between the two GCMs play a large role in determining the outputs of the hydrological models, and conclude that regional impact studies must use more than one global model. They also provide a list of expected changes in local climate in Sweden, including increased runoff in the north and decreased runoff in the south of the country, and changes in seasonal behaviour.

Modelling Feedbacks

The modelling approaches discussed up to this point offer complex representations of the climate system, and can resolve sub-continental (GCMs), and even regional (RCMs), variations in climate with a fair level of accuracy. Furthermore, despite difficulties in predicting certain highly-local climate characteristics like precipitation, there is good reason to believe that improvements both in model resolution and in the representation of physical processes will lead to better models, better agreement with climate observations, and ultimately more reliable predictions. However, climate models currently cannot provide the kind of fine detail necessary for hydrological modelling at the catchment scale, and computational requirements mean that very few studies can be run and for very short simulated periods; the results of simulations can also be very hard to evaluate because of the complexity of these high-resolution models. On top of all these problems, GCMs ignore connections between the climate system and the socioeconomic system, which share important feedbacks – after all, anthropogenic emissions are responsible for the current changes in global climate.

Therefore, another option is the use of more comprehensive, but less complex models. Simple models offer a means of improving our understanding of the feedbacks that connect the climate with the carbon cycle, the energy sector, economics, population, government, and social welfare. These models sacrifice resolution for completeness, and so cannot provide useful inputs to catchment-scale hydrological projects, or include the number of physical processes present in GCMs; however, high-resolution models currently cannot provide reliable regional information either, as demonstrated above. Furthermore, simple models supply a context for climate change, which global models driven by emissions scenarios cannot offer; they run very quickly on simple computers, making multiple simulations and an assessment of model sensitivities possible; and they improve our understanding of the feedbacks between the socio-economic and climate systems.

Both Nordhaus and Boyer and Fiddaman have developed simple climate-economy models, called DICE and FREE, respectively; such models attempt to include a broad range of relevant disciplines in the modelling process, and to provide information suitable for policy development. Our current work builds on DICE and FREE, improving the representation of physical processes by developing more complex models of the climate system and the carbon cycle. We use a methodology called System Dynamics that can both deal with long time delays, multiple feedback processes, and other elements of dynamic complexity; and also combine the very different research fields of engineering and the natural sciences, political science, and economics. Overall, our research will contribute to an understanding of climate change, add to number of approaches available for climate change modelling, introduce system dynamics with its explicit feedback processes to the earth-system modelling community, and link scientific and socio-economic disciplines.

Sulphur Cycle

Dimethyl sulfide (DMS, CH_3SCH_3) is a semi-volatile organic compound produced by several environmental sources, totaling 25–80 TgS in global emissions. Natural sulfur is produced from sources such as volcanoes, biomass burnings, and fumaroles, with the largest contribution coming from phytoplankton. However, the production rates of DMS and its precursor—dimethylsulfoniopropionate (DMPS)—varies a great deal throughout various functional groups and environmental conditions. Marine ecosystems contribute 50–60% of natural sulfur emissions and dominate the sulfur cycle in the Southern Hemisphere. Once released from the mixed layer, the gaseous precursor—DMS—is oxidized in the lower atmosphere, resulting in a series of products, including sulfur dioxide (SO_2). The SO_2 has a final reaction with HO_x oxidants, concluding with sulfate aerosol particles. The sulfur-containing droplets disperse incoming solar radiation and act as condensation nuclei for cloud seeding, modifying the global radiative budget both directly and indirectly. The compound first gained notoriety within the scientific community for its role in the CLAW hypothesis (Charlson, Lovelock, Andreae, and Warren).

An exhaustive anthology of articles exists examining individual impacts of climate change on DMS production. Findings detailed by author—such as Six et al., Flombaum et al., Cameron-Smith et al., and others—have served as an invaluable foundation for the present work. However, to the best of the authors' knowledge, limited comprehensive syntheses have been attempted. This shortage of integrated information motivates our current analysis. The examination began by reviewing the modern portrayal of the sulfur cycle in the reduced climate model Hector. We then built upon the typical biological representation by constructing a dynamic component which accounts for acidification. Next, the location and abundance of several phytoplankton classes were cataloged using present and end of century (EOC) simulations. These values were directly interpreted to quantify dimethyl sulfide contributions from each group. The investigation concluded with a range of sensitivity tests verifying the need for further evaluation of DMS in the global climate system.

In addition to bringing together the disparate effects of climate change on the phytoplankton community, this publication examines changes on a regional scale. We have partitioned global data into three meridional zones: northern high latitude (NHL), low latitude (LL), and southern high latitude (SHL). The segregation of information illustrates the diverse repercussions climate change can have on each section and the consequent influence on natural sulfur emissions (ESN). Ultimately, a localized approach attempts to calculate the varying impact that mounting carbon pollution has on natural sulfur production and, in turn, its influence on the radiative budget. This marine biogeochemical (mBGC) feedback is then contrasted with the familiar greenhouse gas equivalent—anthropogenic carbon dioxide (CO_2)-and its role as a forcing agent.

Our exploration is built upon the modern representation of the sulfur cycle within climate simulations. The Joint Global Change Research Institute (JGCRI) reduced-complexity model Hector was utilized due to its nearly instantaneous rendering-time and comprehensive format, which keeps each section separate and easy to manipulate. The model includes a one-pool atmosphere, land, and ocean component. The ocean is divided into a high- and low-latitude surface, an intermediate, and a deep box. Carbon first enters the high-latitude box and sinks to its depth. It then returns through the intermediate to the low-latitude surface box, simulating a simple thermohaline circulation. In its current state, Hector matches both historical trends and future projections comparable to observations and models included in the most recent Coupled Model Intercomparison Project. Hector is an open source platform and allows for the addition of new components. This duality provided an ideal opportunity to parameterize natural sulfur emissions and evaluate the output decisively. For a more detail description of Hector, refer to Hartin et al. The simple climate model (SCM) includes both a direct and an indirect sulfate aerosol equation (SO_x). Considering the prime objective of the present work is to streamline dominant effects of ESN, we focused on the latter. The formula employed in Hector was adopted from Joos et al.:

$$RF_{SO_x}Indirect = -0.6 \times (ln\frac{ESN + ESO_{xt}}{ESN}).(ln\frac{ESN + ESO_{xt0}}{ESN})^{-1}$$

The calculation requires a fixed input rate for ESN, anthropogenic sulfur emission during the year 2000 ($ESO_x t0$), and a time series of man-made pollutants (ESO_{xt}). The value used for natural sulfur emissions was 42 TgS, matching the flux currently found in Hector. From equation:

$$RF_{SO_x}Indirect = -0.6 \times (ln\frac{ESN + ESO_{xt}}{ESN}).(ln\frac{ESN + ESO_{xt0}}{ESN})^{-1},$$

it is clear that any fluctuation in phytoplankton abundance is not reflected regionally or temporally. It should be noted that in Joos' original report, the leading coefficient for the radiative forcing (RF) value is -0.8 Wm^{-2}, while Hector uses -0.6 Wm^{-2}. This minor discrepancy speaks directly to the inadequate understanding of the extent to which sulfur participates in the planetary energy budget. For consistency's sake, we have chosen to follow the path laid out by Hector and used an RF of -0.6 Wm^{-2} for the remainder of our work.

Natural Sulfur Emissions: Ocean Acidification (ESNa)

In an attempt to examine the DMS feedback contribution to the sulfur cycle, an additional term was added to equation:

$$RF_{SO_x}Indirect = -0.6 \times (ln\frac{ESN + ESO_{xt}}{ESN}).(ln\frac{ESN + ESO_{xt0}}{ESN})^{-1}.$$

Variations in the new expression are intended to demonstrate the effects that ocean acidification may have on the release of ESN. This was accomplished by utilizing the change in dimethyl sulfide emissions presented in S13. In her report, concentrations of DMS are listed for 19th and 21st century scenarios. The data were calculated based on mesocosm experiments studying the impact a diminishing pH value has on sulfur-producing biota. Including S13 findings into the new equation results in a temporally dynamic representation of natural sulfur emissions with respect to ocean acidification (ESNa).

$$RF_{SO_x} Indirect = RF_{ESO} \times (ln\frac{ESN + ESO_{xt}}{ESN}).(ln\frac{ESN + ESO_{xt0}}{ESN})^{-1}$$

$$+ RF_{ESN} \times (ln \frac{ESN_{xt65} + \Delta ESN_{xt}}{ESN_{xt65}}).(ln \frac{ESN_{xt65} + \Delta ESN_{xt100}}{ESN_{xt65}})^{-1}$$

Beginning with anthropogenic and natural RF terms (RF_{ESO}, RF_{ESN}), these variables are responsible for additional forcing attributable to their respective sources on both a global and a regional scale. Imitating the format used in equation,

$$RF_{SO_x} Indirect = -0.6 \times (ln \frac{ESN + ESO_{xt}}{ESN}).(ln \frac{ESN + ESO_{xt0}}{ESN})^{-1},$$

the log functions were retained in the supplemental expression to simulate saturation of DMS in the atmospheric system. Finally, values derived to complete the new terms— ESN in the year 1865 (ESN_{xt65}) and change in natural emissions due to ocean acidification since 1865 for each year (ΔESN_{xt}) and by EOC (ΔESN_{xt100})—are based on findings reported in S13 and explored in greater depths in successive subsections.

Radiative Forcing: Anthropogenic and ESNa

Equation,

$$RF_{SO_x} Indirect = RF_{ESO} \times (ln \frac{ESN + ESO_{xt}}{ESN}).(ln \frac{ESN + ESO_{xt0}}{ESN})^{-1}$$

$$+ RF_{ESN} \times (ln \frac{ESN_{xt65} + \Delta ESN_{xt}}{ESN_{xt65}}).(ln \frac{ESN_{xt65} + \Delta ESN_{xt100}}{ESN_{xt65}})^{-1}$$

requires RF values for both natural and anthropogenic sulfur emissions on a global and regional scale. While a planetary anthropogenic sulfur emissions (ESO_x) forcing term is supplied by Hector (-0.6 Wm^{-2}), the forcing value needed to be additionally divided in order to determine local variations. A useful table displayed in Smith et al. tabulated SO_2 emissions by country. This enabled a geographical restructuring of pollutants into latitudinal ranges designated for our work—SHL (90–30°S), LL (30–30 °N), NHL (30–90 °N). To continue a strict methodology of assigning sulfur dioxide emissions, records attributed to "International Shipping" were excluded from local divisions, as they did not have a clear point of origin and produced a negligible flux on a regional scale. However, the vessel's contributions were reintegrated upon calculation of a global average. After calculating man-made emissions derived within each hemispheric band, percentage totals were easily evaluated, yielding 5%, 22%, and 64% of the global total for the SHL, LL, and NHL, respectively. These allotments were applied to the planetary average forcing provided by Hector (-0.6 Wm^{-2}), producing values of -0.11, -0.54, and -1.57 Wm^{-2}.

The approach employed for computing RF_{ESN} data could be streamlined due to an image in S13. The image presents change in radiative forcing (Wm^{-2}) from 1865–1874 to 2090–2099 under varying pH-sensitivity scenarios—reference, low, medium, and high situations—ranging from 8.3 to 7.6 pH units. We extracted data from this image and calculated the difference between the high and reference scenarios for the end of century. We utilized this quantity since the intention of our publication is to isolate and observe the effect climate change may have on marine biota and subsequent feedbacks. The global forcing values computed from the image in S13 were then averaged throughout each regional band and across the entire planet: 0.96, 0.40, 0.50, and 0.62 Wm−2, respectively. A list of both ESNa and ESO_x forcing values can be found in table.

Location	ESO_x (Tg)	Percent Total (ESO_x)	$RF_{ESO}X$ .Wm−2 Σ	RF_{ESN}.Wm−2 Σ
Southern High Latitude	4.830	4.52%	-0.110	0.963
Low latitude	23.592	22.08%	-0.538	0.395
Northern high latitude	68.668	64.25%	-1.567	0.498
International shipping	9.779	9.15%	-0.223	-
Global total	106.869	9.15%	-0.610	0.615

Anthropogenic Sulfur Emissions

To create a complete anthropogenic sulfur emissions dataset, we expanded the ESO_x table described in the previous subsection. For the purpose of calculating an RF term, we focused on pollution from the year 2000. However, Smith et al. only included historical data for the start of each decade—1910, 1920, etc.—between 1850 and 2000. In order to compute corresponding values for intermediate years, a simple linear regression formula was applied. The periods succeeding the millennium required a companion source of information. The necessary data were found in van Vuuren et al. describing various representative concentration pathway (RCP) scenarios of human pollutants throughout the 21st century. Since findings from the high pH-sensitivity scenario were used to calculate a radiative forcing value for ESNa, we elected to apply projections under an RCP8.5 future to preserve consistency. Under this pathway, a global average value was provided for each year following 2000. The global average was subsequently divided into the three regional bands using the percentages established for RF computations.

Natural Sulfur Emissions

At present, there is a shortage of dimethyl sulfide emissions information for both the contemporary era and throughout the coming century. Taking this into consideration, in order to create our ESNa dataset, we again relied on a linear regression formula. First, data for the slope was provided by S13. One of their figures displays the meridional "Change in DMS Flux" from 1865–1874 to 2090–2099. Next, it was necessary to find emissions by latitude for a single year within the prescribed timeline to act as our reference point. The required statistics have been provided in Simó and Dachs, 2002. By combining these values, we were able to derive ESNa for each latitude in 1865—the preliminary year chosen by S13—and 2100. As a verification of these methods, natural sulfur emissions in 1865 and 2100 were integrated, giving 29 and 22 TgS year–1, respectively, successfully matching data reported in S13.

A final alteration applied to the data-stream was to impose a "bend" in the estimated slope. This was done to properly reflect the changes of a warming climate as experienced throughout the ocean. Since general consequences of global warming were not readily measurable until the mid-20th century, change in DMS emissions was maintained at 0 until 1950. The inflection point was selected to represent the observed decline of seawater density found by all major Earth system models (ESM), marking the beginning of the measurable impact of climate change on the planet. Following the middle of the century, we imposed a constant slope yielding ESNa values until the year 2100. With all of the required information for the linear regression formula, the remaining values of the ESNa dataset were calculated.

Community Shifts (ESNc)

Alteration in abundance and habitat of phytoplankton is the second effect explored in our report. We have defined phytoplankton as a single functional group. However, the remainder of our analysis tested changes which have varying outcomes for the respective classes and required further enumeration. In this context, phytoplankton were organized according to their distinct concentration of the organic sulfur precursor—DMSP. The following portion of our work aims to isolate the major biological forms and to examine the ways in which they are likely to respond to impending changes in climate. The phytoplankton were cataloged as non-DMS producers (cyanobacteria and diatoms) and DMS producers (Phaeocystis, coccolithophorids, along with a dinoflagellate et al. group). By contrast with ocean acidification, this analysis was not accomplished by developing an equation, but rather via an exhaustive synthesis of established data from the beginning and end of century.We evaluated variations attributed to nutrient stress, rising sea surface temperature (SST), and increasing exposure to solar irradiance. In keeping with the tone of the "Experiments", phytoplankton have been ordered in relation to the data collection methodology, rather than DMS production rates.

Cyanobacteria and Coccolithophorids

The primary source for contemporary phytoplankton distributions is Gregg et al., 2003. In their work, ESM simulations of chlorophyll abundance attributed to cyanobacteria and coccolithophorids are computed for February and June. These results were averaged to derive an annual value for each latitudinal band. Approximations for coccolithophorids were then correlated with the chance of finding Emiliania huxleyi (EHUX) throughout the global hydrosphere at present and EOC. Using an elementary relation, we calculated a future-to-modern ratio for the probability of EHUX presence and multiplied the result with the current chlorophyll concentrations to tabulate end of century values.

To analyze community shifts for cyanobacteria, data derived from F13 were used. Two picocyanobacterial genera—Synechococcus and Prochlorococcus—demonstrate the growth of cyanobacteria, since they outcompete and supersede their DMS producing counterparts. Flombaum's group measured the progression of cell abundance, allowing us to establish a percentage change. A minor adjustment was made, since changes reported by F13 were derived from an RCP4.5 scenario. Under these parameters, the ocean would simply warm on average by 1.4 °C relative to 5.5 °C from an RCP8.5 future. Operating with outputs from Flombaum and the 4.5 pathway temperature increase, we used a slope to calculate cell count growth under the more severe 8.5 scenario while assuming no differential adaptation to the higher temperatures. Ultimately, these changes were applied to modern values, deriving future concentrations at each latitude.

Phaeocystis

Gathering statistics pertaining to Phaeocystis proved to be the most formidable obstacle. Unlike other species, a great deal of effort was spent trying to find any habitat or concentration data for the foam-producing algae. Unfortunately, we were unsuccessful in obtaining any clear figures, measurements, or other information, and ultimately used an image included in Vogt et al. reporting cell concentrations from limited depth-resolved stations. Remote sample locations generated extreme sensitivity for the strong DMS emitter. A trivial adjustment to the extrapolated data could significantly influence the overall flux of natural sulfur. Similar to the coccolithophorids calculation, these values were multiplied with a future-to-modern ratio of DMS concentrations associated with the phytoplankton in question, deriving EOC estimations.

Diatoms and Dinoflagellates

Statistics of diatom and small phytoplankton abundance were taken from Marinov et al. Their publication provided an itemized listing of biomass values for diatoms in conjunction with habitat distribution at the beginning and end of century. Information involving small phytoplankton led to the construction of a "catch-all" category for any overflow marine biota which contribute to ESN. As a concession to the finite number of simulations possible, we argue that this group is predominantly populated by dinoflagellates and that their DMS flux would reflect this. Concentrations were collected from diagrams in M13 illustrating small phytoplankton biomass at present and by the year 2100. Quantities associated with the remaining microalgal types—cyanobacteria, Phaeocystis, and EHUX—were then removed from the data, isolating the intended class.

Normalization

As a verification of the data compiled for each functional group, our findings were normalized relative to a table provided in M13. Included in this chart were concentrations for diatoms, small phytoplankton, and total sea surface biota. However, since values for diatoms were taken directly from Marinov's index, it was not subject to refinement. An added complication to the process arose as each group was subdivided into imprecise biome regions such as Equatorial or Subtropical Northern Hemisphere. While aided by an image representing the author's intent for these habitats, we had to use expert judgment when assigning various locations into distinct latitudinal zones. This was not possible in two meridional portions to the north (40–50 °N and 60–70 °N). Within these settings, no clear biome dominated the bands and, instead, a linearization from the proximal ecosystem categories was imposed.

Once arranged into appropriate domains, estimates from the methods explored in preceding subsections were refined relative to measurements found in M13. We first calculated the ratio of small to total plankton within each biome. These percentages provided a scale, which was then used to adjust our phytoplankton values to produce a total concentration equivalent to that reported by Marinov. As a supplement to the process, since findings by Vogt et al. had a deficiency of Phaeocystis data at SHL, we scaled results between 60–80°S to represent the established presence of the predominant DMS producer.

Radiative Forcing: Community Shifts

The final step for assessing outcomes from planktonic community shifts was to relate marine biotic concentrations to their corresponding impact on the global radiative budget. Beginning with a few simple conversions, values from Elliott provided average N:S ratios, which were used to calculate sulfur concentrations (C_j) for each functional group. Additionally, this report included percentage yields (Y) for dimethyl sulfide from its precursor—DMSP. These quantities were then substituted into the following expression:

$$Source_{DMS} = g.Z.(\frac{C_j}{k_3}).Y$$

Equation $Source_{DMS} = g.Z.(\frac{C_j}{k_3}).Y$ was adopted and adapted from a grazing rate equation found in Sarmiento et al. The calculation determines the frequency with which DMSP is released via predation and the consequent formation of DMS. In this context, (g) is a shorthand for maximum growth rate and (K_3) is a stand-in for half saturation ingestion. These constants were assigned to be 1 day–1 and 1 mmolN m^{-3} respectively. Finally, (Z) denotes zooplankton concentrations which were listed in M13.

The next step in the process was to quantify a DMS time constant:

$$Time\,Constant_{DMS} = (k_B.(0.1.(N_p)^{0.5}))^{-1}$$

In the simple Equation:

$$Time\,Constant_{DMS} = (k_B.(0.1.(N_p)^{0.5}))^{-1},$$

(kB) is the bacterial kinetic coefficient, assumed to be 30 (day mmolN m−3)−1. The second term represents concentration of bacteria, which is itself a function of phytoplanktonic nitrogen (Np). Net dimethyl sulfide flux was then found by multiplying frequency of production (Equation), the DMS time constant (Equation), and the appropriate piston velocity. In order to match accepted dimethyl sulfide emissions, results had to be scaled by a factor of 5. Returning to S13, latitudinal correlations were evaluated using the data reported in the figures "Change in DMS Flux" and "Radiative Forcing". These associations were used to convert our calculated emission rates to RF values, both regionally and globally. Finally, the calculated changes in radiative forcing were added, offline, to the original anthropogenic emissions to estimate the total sulfur cycle with respect to phytoplanktonic community shifts.

CO$_2$ mBGC Feedback

One of the goals of the present work is to demonstrate the influence marine ESN have on the global radiative budget, both historically and infuture decades. In order to give context to our findings, results we recompared to the forcing agent most familiar throughout the scientific community—anthropogenic carbon dioxide. This requires an examination of the feedback effect CO_2 and marine biota have on one another. Unfortunately, Hector itself does not include a representation of the biological pump at this time and could not be used for our intended purposes. Instead, one of the authors (CH) provided a simplified model of the carbon ocean component from Hector in an Excel format. This allowed us to easily add in equations and track the flow of carbon throughout the ocean. The document served as a skeletal structure, which we developed to include an atmosphere in addition to a solubility and biological pump. Initial values for the various reservoirs were taken from Watson and Liss, and the solubility pump is based on the formula outlined in Sarmiento and Gruber. The inclusion of the solubility pump required represent ation of the carbon cycle chemistry as a function of time in both surface ocean reservoirs. The model was ultimately parameterized to ensure this flux paralleled historical and projected concentrations to represent the changes to climate, including ocean acidification, accurately. Forcings attributed to CO_2 emissions were then calculated using the expression found in Hector:

$$RF_{CO_2} = 5.35 \times ln\frac{Ca}{C0}$$

Equation $RF_{CO_2} = 5.35 \times ln\frac{Ca}{C0}$ assigns 5.35 Wm^{-2} as a scaling factor in addition to requiring both the initial and current atmospheric concentrations (Ca, Co). The starting value was set at 278 ppm.

Biological Pump

Within our Excel-Hector, the biological pump is represented using two formulas, both of which are a function of annual mean surface nitrate concentrations (N):

$$\text{Biological Pump}_{surface} = R \cdot N \cdot (1 - Rm)$$

$$\text{Biological Pump}_{Int,Deep} = R \cdot N \cdot (Rm)$$

In equations:

Biological Pump $_{surface} = R \cdot N \cdot (1 - Rm)$ and Biological Pump$_{Int,Deep} = R \cdot N \cdot (Rm)$, (R) is the standard Redfield ratio used to convert the controlling element into carbon values and (Rm) signifies the remineralization percentage, which determines the amount of nitrate that will remain at the respective depths. Equation Biological Pump $_{surface} = R \cdot N \cdot (1 - Rm)$ is subtracted from reservoirs calculated in the mixed layer at both high and low latitudes, while equation Biological Pump$_{Int,Deep} = R \cdot N \cdot (Rm)$ is added to the intermediate and deep equivalent, ensuring both conservation of nitrogen atoms and the preservation of nitrate abundance. The utilization of both expressions allows for control of the biological sequestration rate, which we vary throughout the present century.

To ensure accuracy of the model, inputs were set to calculate: a biological pump strength of −120 ppm for the present day; and an additional 30 ppm in the atmosphere by EOC relative to a climate change free reference run. It should be noted that the decrease in biopump efficiency attributed to global warming is greater than that reported in analogous models. This exaggerated difference was chosen to demonstrate the potential significance of DMS relative to an extreme baseline. Nitrate values were held constant from initialization of the model through 1950, and then follow a linear decay to endpoints prescribed during parameterization. The midcentury date was chosen to parallel our DMS examination.

Sensitivity Tests (ESNuc and CO_2uc)

Limited available data proved to be a continual challenge throughout this investigation. As such, findings from the present work were subjected to a series of sensitivity tests to calculate ranges of uncertainty. Each assessment was conducted separately: an adjustment was made to our results, the corresponding output was calculated, and the data were reverted to the original state before the analysis continued.

Uncertainty: Ocean Acidification First, we reevaluated claims made in S13. Considering our experiment utilized their high acidification scenario, an exercise was designed to determine results for a low-end outcome. The test focused on the fact that, while diminishing pH levels will affect all marine biota, calcium shells surrounding coccolithophores result in a greater susceptibility for their community. Therefore, we computed the ratio of

EHUX to total phytoplankton concentrations and multiplied the result with the change in radiative forcings by EOC. This effectively isolated the impact of coccolithophores, which was then subtracted from our ΔRF to determine the extent of uncertainty.

Uncertainty: Community Shifts

To verify our community shift findings, we designed a test for each of the three meridional zones. The low latitude assessment focused on F13's picocyanobacterial data derived from an RCP4.5 scenario. As previously stated, in order to keep EOC outcomes consistent throughout this investigation,we had to scale Flombaum's output to match an RCP8.5 temperature increase. Therefore, we decided it was appropriate to calculate the difference between our results and the anticipated changes reported by F13.

For both high-latitude regions, the greatest source of uncertainty came from contemporary Phaeocystis data found in Vogt et al. In this particular report, it is explained that measurements were taken from remote stations, resulting in a thorough analysis of the Arctic Ocean and an underrepresentation of its southern counterpart. Unfortunately, alterations could not be made directly to Vogt's findings due to subsequent modifications of the data. Instead, the normalized small phytoplankton biomass was: cut in half between 50–60 °N; and scaled an additional 10 times in the Southern Hemisphere ice biome to match accepted annual DMS emissions. Although the latter adjustment may appear to be excessive, we believe it further validates the necessity for additional DMS investigations. Throughout the present work, we were limited to a single report regarding Phaeocystis, resulting in striking modifications, such as those used to produce acceptable statistics.

An additional test was conducted in the SHL, focusing on the projected change in Phaeocystis abundance. Estimated data from model output used for our analysis reflect approximately a 50% decrease in Phaeocystis by EOC. However, in the current literature, there are also publications which suggest a pole ward migration of the genus. Considering these potentially contrasting scenarios, we changed the decline in concentration to 10%. Finally, outputs from all three sensitivity tests in addition to the reference run were compared, and both the high and low values were determined, producing our uncertainty ranges.

Uncertainty: CO_2

Currently, a considerable amount of scientific and Earth system modeling attention is given to reducing the uncertainty associated with anthropogenic CO_2. This level of refinement is in contrast with the present work, which demonstrates the many improvements still necessary for contemporary or projected ESN data. To quantify this discrepancy, we reviewed the current literature to determine an acceptable range of warming attributed to CO_2 and a waning biological pump. According to the fifth Coupled Model Intercomparison Project, by the end of century, atmospheric concentrations are expected to increase due to feedback by 50–100 ppm, with <20% attributed to the global aquatic ecosystem. This suggests an increase of 10–20 ppm from the carbon dioxide mBGC interaction. Excel-Hector was reparametrized with both a 10 and 20 ppm change in biopump strength, relative to the reference EOC value, and rerun to calculate the respective ΔRF for the accepted range.

Anthropogenic Sulfur Emissions (ESOx)

In its present configuration, Hector uses equation,

$$RF_{SOx}Indirect = -0.6 \times (ln\frac{ESN + ESO_{xt}}{ESN}).(ln\frac{ESN + ESO_{xt0}}{ESN})^{-1}$$

to quantify anthropogenic sulfur pollutants while holding natural emissions constant. Throughout the modern century, both local and global ESOx share characteristic trends to varying degrees. Between 1850 and 1900, the forcings begin to decrease, resulting in negative RF values, reaching a minimum between 1970 and 2000. Following these inflection points, under an RCP8.5 scenario, all outcomes begin and continue to rise until EOC. This increase can be attributed to a growing effort from industrialized nations to reduce their contribution of sulfur dioxide to climate. We observe the greatest decline in RF at northern high latitudes, since they host a majority of polluting nations. This region reaches a minimum value of –1.58 Wm–2 in 1970. The negative radiative forcing reduces the global average, causing the planetary mean to have the second-lowest RF value in the plot. Worldwide irradiance reaches–0.6 Wm–2 in the year 2000, suggesting equation,

$$RF_{SOx}Indirect = -0.6 \times (ln\frac{ESN + ESO_{xt}}{ESN}).(ln\frac{ESN + ESO_{xt0}}{ESN})^{-1}$$

is performing as anticipated. The low-latitude outcome follows the same characteristic pattern but on a mitigated scale, reflecting the smaller number of developed countries. Finally, the Southern Ocean experiences a minimal decline—attributable mainly to South Africa and Australia—before stabilizing near 0 Wm⁻².

Natural Sulfur Emissions: Ocean Acidification (Esna)

Figure isolates our addition to the standard indirect sulfate aerosol equation. Adopting the methods and data reported in S13, linear slopes are imposed. Starting in the 1950s, each function begins to show the impact of ocean acidification. Due to the high phytoplanktonic concentration in the Southern Ocean, the largest rise in forcing is observed at SHL, reaching a value of 0.96 Wm–2 by EOC. Analogous to ESOx, this large local growth was the dominant factor governing the global average. The planetary mean has the second-highest value among the RF at 0.62 Wm⁻², representing a possible impact 4 times that of marine carbon cycle feedback.

Ultimately, equation,

$$RF_{SO_x} Indirect = RF_{ESO} \times (ln \frac{ESN + ESO_{xt}}{ESN}).(ln \frac{ESN + ESO_{xt0}}{ESN})^{-1}$$

$$+ RF_{ESN} \times (ln \frac{ESN_{xt65} + \Delta ESN_{xt}}{ESN_{xt65}}).(ln \frac{ESN_{xt65} + \Delta ESN_{xt100}}{ESN_{xt65}})^{-1}$$

characterizing both man-made and natural sulfur emissions—was used in Hector to compute values shown in figure. The effect of a dynamic ESN component is evident, since the increase in radiative forcing substantially shifts end of century values. Introducing DMS emissions effectively reversed the sign in all regional and global divisions, barring northern high latitude. These combined components result in a planetary average of 0.42 Wm⁻².

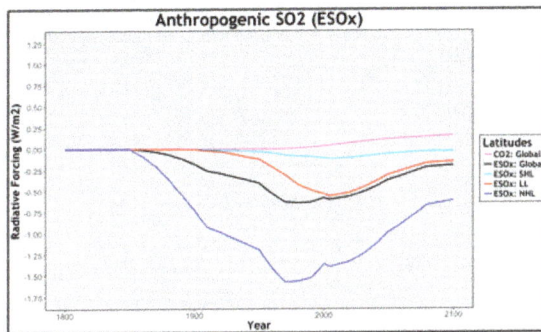

(a)

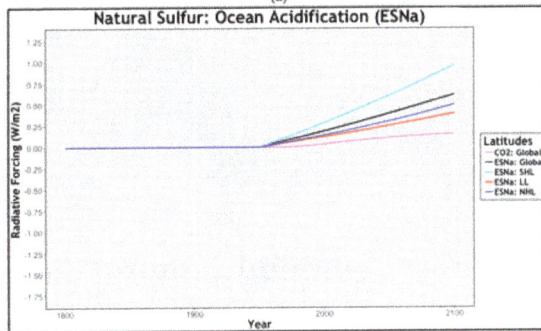

(b)

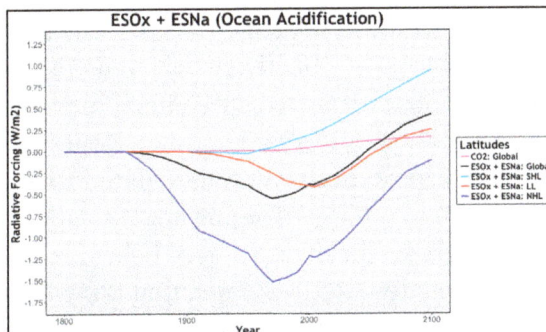

(c)

Impacts the sulfur and CO_2 marine biogeochemical (mBGC) feedbacks have on the global radiative forcing budget, from 1800 to 2100. Included in the illustration are CO_2 global impact (pink), global average sulfur impact (black), southern high-latitude sulfur impact (90–30 °S) (light blue), low-latitude sulfur impact (30–30 °N) (red), and northern high latitude sulfur impact (30–90 °N) (blue): (a) Anthropogenic SO_2 and CO_2 forcings; (b) Natural influence of sulfur with respect to ocean acidification; (c) Natural and anthropogenic sulfur forcings taken together while considering acidification; (d) Natural influence of sulfur with respect to community shifts; (e) Natural and anthropogenic sulfur forcings accounting for phytoplankton community shifts.

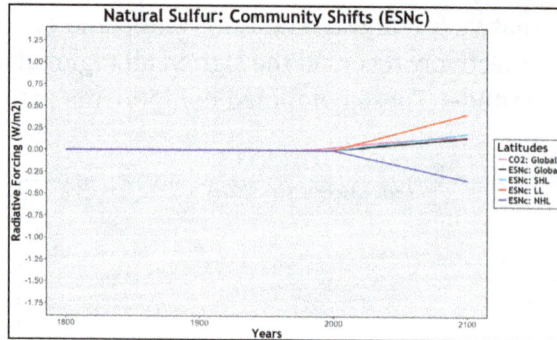

(d)

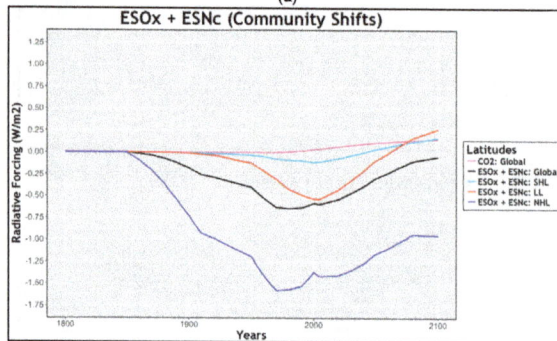

(e)

Natural Sulfur Emissions: Community Shifts (ESNc)

Distinct from the changes explored in the previous subsection, shifts in community structure result in both intensification and reduction to the RF. As seen in Figure, northern high and low latitudes exhibit opposite trends in the 21st century. By the year 2100, LL values approach 0.42 Wm–2, effectively balancing the decline to −0.34 Wm⁻² at NHL on the global average. An unexpected outcome of biogeographic evolution was observed in the south. Since ecosystems in the SHL experience environmental changes similar to the northern counterpart, both were anticipated to adjust in a comparable manner. However, the increase in radiative forcing at SHL suggests a loss of DMS-producing biota. A working hypothesis for the divergent behavior is again associated with limited Phaeocystis records. The few stations collecting Phaeocystis data near Antarctica develop a disproportionate representation between 60 °S and 80 °S. Therefore, similar percentage losses experienced near both poles lead to a greater decline in the southern population and, by extension, DMS emission rates.

A final community shift graph was produced representing the collective anthropogenic and natural sulfur forcings. The new trends show phytoplankton relocation countering man-made emissions in both low and southern regions, leading to a positive radiative forcing by 2100. However, in NHL, changes in the marine biota have an amplifying effect on the sulphur pollutant. These contrasting outcomes nearly balance one another, reducing any changes from a global perspective.

Change in Total Chlorophyll Concentration: (RCP8.5–Contemporary)

An alternate representation of community shifts shows the change in chlorophyll concentration between end-of-century and modern-day simulations. Established data of phytoplanktonic classes were synthesized and organized into two groups: DMS producers (Phaeocystis, EHUX, dino flagellates et al.) and non-DMS producers (cyanobacteria, diatoms). Figure illustrates the results reported in table, since the latitudinal trends follow qualitative projections in the current body of literature.

Next to Antarctica, there is an increase in DMS producers. Although this seems counterintuitive relative to the results reported in the preceding subsection, our projected growth for the dinoflagellate group outweighs any loss of Phaeocystis. Ultimately, the shift to a somewhat weaker producer will reduce DMS emissions but increase the total chlorophyll concentration. Between 60 °S and 30 °S, there is a migration poleward of DMS-producing plankton and consequent replacement by diatoms. Low latitudes show a growing dominance of cyanobacteria under reduced nitrate abundance and rising SST. These small algae supersede the strong dimethyl sulfide-emitting plankton. An increase in non-DMS producers reaches a peak at 25°, representing maximal expansion of cyanobacteria and minimal reduction to the diatom population. Although the equatorial band is dominated by Synechococcus and Prochlorococcus, the reisnotan equivalent maximum at the Inter tropical Convergence Zone (ITCZ), as reported in F13. This is a consequence of the normalization process. Finally, in NHL, there is a small decline in diatom abundance with a trough at 65 °N and a concurrent growth of the dinoflagellate et al. group, which stabilizes moving toward the pole.

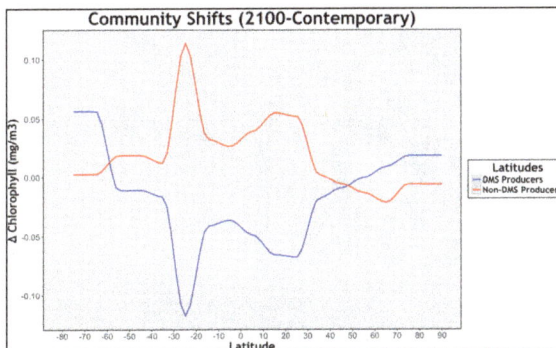

Change in phytoplankton concentration (mg Chl. m⁻³) at each latitude between the end
of century (under an RCP8.5 scenario) and the modern day. Included are non-DMS
producers (cyanobacteria, diatoms) (red) and DMS producers (Phaeocystis, coccolithophores, dinoflagellate et al. group) (blue).

Histograms

This type of chart was chosen for Figure to allow for direct comparison to radiative forcings plots reported in external publications. Both of the figures begin with changes in RF
due to the anthropogenic CO_2 mBGC feedback and SO_2 pollutants. Next are the changes
in forcing attributed to dynamic ESN with respect to ocean acidification and community
shifts—Figure, respectively. Finally, the natural and man-made combined influence on the
global radiative budget is shown. Positioned directly to the right of the dividing line there
are visual representations of the sensitivity tests, uncertainty for global natural sulfur emissions with respect to acidification is more than six times that of CO_2 mBGC feedback. This
demonstrates the need for further quantitatively driven analyses, and for refinement of the
current knowledge base. Results for uncertainty in ESN while analyzing community shifts
are smaller, and the global average is comparable to that of carbon dioxide. In both figures,
the greatest source of uncertainty is in the prime habitat of phytoplankton, the Southern
Ocean. It should be noted that values used for these plots are from EOC and lie beyond the
RF troughs reported from anthropogenic emmisions at the start of the century.

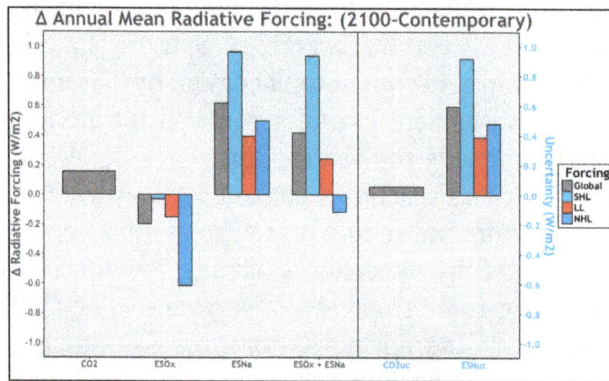

(a)

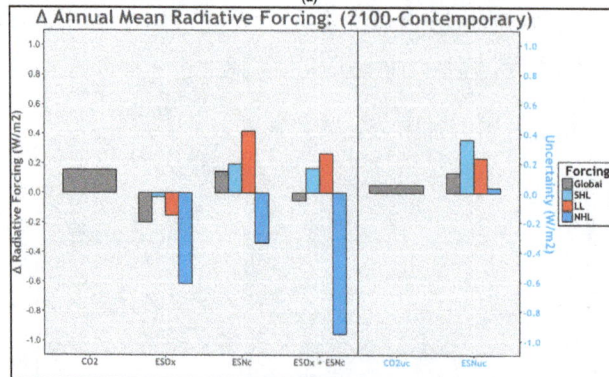

(b)

Results from the investigation are shown in histogram format. Representative global (gray) and regional changes—southern high latitude (SHL: 90–30 °S) (light blue), low latitude (LL: 30–30 °N) (red), northern high latitude (NHL: 30–90 °N) (blue). Included in order: the change in radiative forcing from CO_2 mBGC feedback (CO_2), anthropogenic SO_2 (ESO_x), natural sulfur (ESN), combined man-made and natural sulfur ($ESO_x + ESN$), CO_2 uncertainty (CO_{2uc}), and natural sulfur uncertainty (ESNuc): (a) Results with respect to ocean acidification; (b) Findings from our examination of planktonic community shifts.

Nitrogen Cycle

Nitrogen (N) is an essential component of DNA, RNA, and proteins, the building blocks of life. All organisms require nitrogen to live and grow. Although the majority of the air we breathe is N_2, most of the nitrogen in the atmosphere is unavailable for use by organisms. This is because the strong triple bond between the N atoms in N_2 molecules makes it relatively unreactive. However organisms need reactive nitrogen to be able to incorporate it into cells. In order for plants and animals to be able to use nitrogen, N_2 gas must first be converted to more a chemically available form such as ammonium (NH_{4+}), nitrate (NO_3), or organic nitrogen (e.g. urea - $(NH_2)2CO$). The inert nature of N_2 means that biologically available nitrogen is often in short supply in natural ecosystems, limiting plant growth.

Nitrogen is an incredibly adaptable element, existing in both inorganic and organic forms as well as many different oxidation states. The movement of nitrogen between the atmosphere, biosphere, and geosphere in different forms is called the nitrogen cycle, one of the major biogeochemical cycles. Similar to the carbon cycle, the nitrogen cycle consists of various reservoirs of nitrogen and processes by which those reservoirs exchange nitrogen.

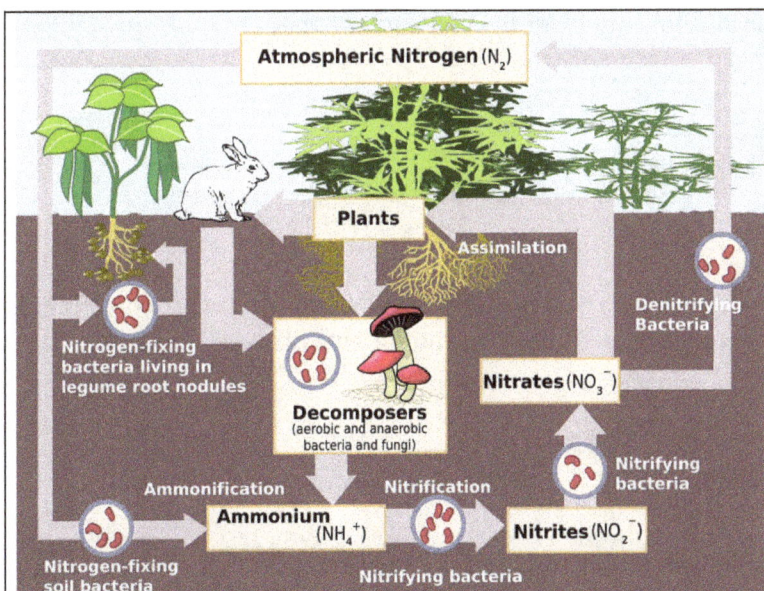

Steps in the Nitrogen Cycle

1. Atmospheric Nitrogen Fixation: Lightning breaks nitrogen molecules (N_2) apart and combines them with oxygen (O_2) to form nitrogen oxides (N_2O) or with hydrogen (H) to form ammonia (NH_3). Nitrogen oxides dissolve in rain forming nitrates (NO_3). Nitrates are carried to the ground with the rain.

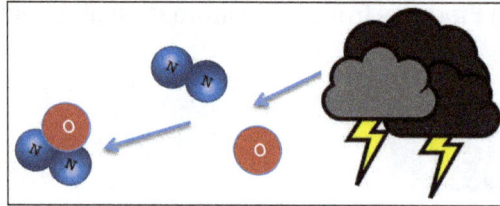

2. Nitrification: Most atmospheric nitrogen is "fixed" (made useable) and changed to ammonia (NH_3) by bacteria in the soil and attached to the roots of legumes (peas, beans, peanuts and soybeans). A few plants can use ammonia but most cannot. Through the process of nitrification, this problem is solved. Nitrifying bacteria change the ammonia in the soil to nitrites and then into nitrates. The nitrates are dissolved in water and absorbed through the roots of plants.

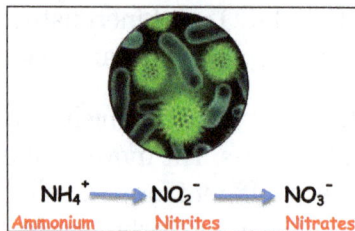

$$NH_4^+ \longrightarrow NO_2^- \longrightarrow NO_3^-$$
Ammonium   Nitrites   Nitrates

3. Assimilation: Assimilation in the process whereby plants absorb the nitrates and/or ammonium from the soil and use them to make proteins.

Nitrates   Ammonium

4. Ammonification: Decomposers (fungi and bacteria) convert the remains of dead plants and animals to ammonia plus other substances.

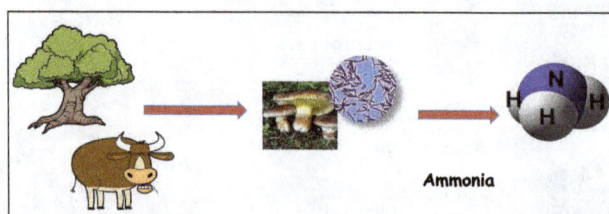

Ammonia

5. Denitrification: Nitrification occurs again, changing the ammonia back into nitrites and then nitrates. Denitrifying bacteria in the soil then change the nitrates into nitrogen gas (N_2), which is released back into the atmosphere to start the cycle again.

Human Alteration of the Nitrogen Cycle

Human activities continue to increase the amount of nitrogen cycling between the living world and the soil, water, and atmosphere. This human driven global change is having serious impacts on ecosystems around the world because nitrogen is essential to living organisms and its availability plays a crucial role in the organization and functioning of the world's ecosystems. In many ecosystems on land and sea, the supply of nitrogen is a key factor controlling the nature and diversity of plant life, the population dynamics of both grazing animals and their predators, and vital ecological processes such as plant productivity and the cycling of carbon and soil minerals. This is true not only in wild or unmanaged systems but in most croplands and forestry plantations as well. Excessive nitrogen can pollute ecosystems and alter both their ecological functioning and the living communities they support.

Most of the human activities responsible for the increase in global nitrogen are local in scale, from the production and use of nitrogen fertilizers to the burning of fossil fuels in automobiles, power generation plants, and industries. However, human activities have not only increased the supply, but also boosted the global movement of various forms of nitrogen through air and water. Because of this increased mobility, excess nitrogen from human activities has serious and long-term environmental consequences for large regions of the Earth.

The impacts of human interference in the nitrogen cycle that have been identified with certainty include:

*   Increased global concentrations of nitrous oxide (N_2O), a potent greenhouse gas, in the atmosphere as well as increased regional concentrations of other oxides of nitrogen (including nitric oxide, NO) that drive the formation of photochemical smog;

*   Loss of soil nutrients that are essential for long-term soil fertility;

- Acidification of soils and of the waters of streams and lakes in several regions;

- Increased transport of nitrogen by rivers into estuaries and coastal waters where it is a major pollutant.

Nitrous Oxide and Climate Change

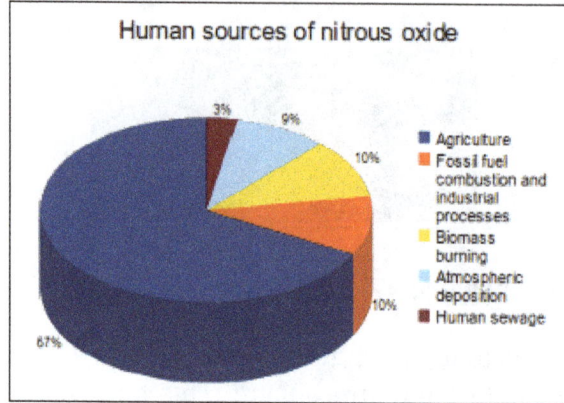

In 2012, nitrous oxide (N_2O) accounted for about 6% of all U.S. greenhouse gas emissions from human activities. Nitrous oxide is naturally present in the atmosphere as part of the Earth's nitrogen cycle, and has a variety of natural sources. However, human activities such as agriculture, fossil fuel combustion, wastewater management, and industrial processes are increasing the amount of N_2O in the atmosphere. Nitrous oxide molecules stay in the atmosphere for an average of 120 years before being removed by a sink or destroyed through chemical reactions. The impact of 1 pound of N_2O on warming the atmosphere is over 300 times that of 1 pound of carbon dioxide.

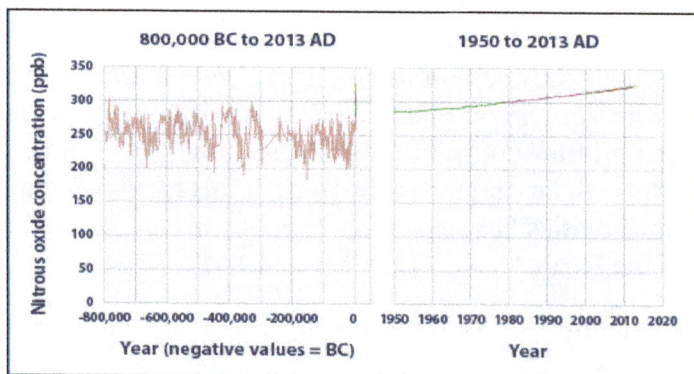

Global concentrations of nitrous oxide over time.

Globally, about 40% of total N_2O emissions come from human activities. Nitrous oxide is emitted from agriculture, transportation, and industry activities, described below:

- Agriculture: Nitrous oxide is emitted when people add nitrogen to the soil through the use of synthetic fertilizers. Agricultural soil management is the largest source of N_2O emissions in the United States, accounting for about 75% of total U.S. N_2O emissions in 2012. Nitrous oxide is also emitted during the breakdown of nitrogen in livestock manure and urine, which contributed to 4% of N_2O emissions in 2012.

- Transportation: Nitrous oxide is emitted when transportation fuels are burned. Motor vehicles, including passenger cars and trucks, are the primary source of N_2O emissions from transportation. The amount of N_2O emitted from transportation depends on the type of fuel and vehicle technology, maintenance, and operating practices.

- Industry: Nitrous oxide is generated as a by-product during the production of nitric acid, which is used to make synthetic commercial fertilizer, and in the production of adipic acid, which is used to make fibers, like nylon, and other synthetic products.

Nitrous oxide (N_2O) emissions in the United States have increased by about 3% between 1990 and 2012. This increase in emissions is due in part to annual variation in agricultural soil emissions and an increase in emissions from the electric power sector. Nitrous oxide emissions from agricultural soils have varied during this period and were about 9% higher in 2012 than in 1990. N_2O emissions are projected to increase by 5% between 2005 and 2020, driven largely by increases in emissions from agricultural activities.

Nitrous oxide emissions occur naturally through many sources associated with the nitrogen cycle, which is the natural circulation of nitrogen among the atmosphere, plants, animals, and microorganisms that live in soil and water. Nitrogen takes on a variety of chemical forms throughout the nitrogen cycle, including N_2O. Natural emissions of N_2O are mainly from bacteria breaking down nitrogen in soils and the oceans. Nitrous oxide is removed from the atmosphere when it is absorbed by certain types of bacteria or destroyed by ultraviolet radiation or chemical reactions.

Within the last century, humans have become as important a source of fixed nitrogen as all natural sources combined. Burning fossil fuels, using synthetic nitrogen fertilizers and cultivation of legumes all fix nitrogen. Through these activities, humans have more than doubled the amount of fixed nitrogen that is pumped into the biosphere every year.

Phosphorus Cycle

Phosphorus is an important element for all forms of life. As phosphate (PO4), it makes up an important part of the structural framework that holds DNA and RNA together. Phosphates are also a critical component of ATP—the cellular energy carrier—as they serve as an energy? Release' for organisms to use in building proteins or contacting muscles. Like calcium, phosphorus is important to vertebrates; in the human body, 80% of phosphorous is found in teeth and bones.

The phosphorus cycle differs from the other major biogeochemical cycles in that it does not include a gas phase; although small amounts of phosphoric acid (H_3PO_4) may make their way into the atmosphere, contributing—in some cases—to acid rain. The water, carbon, nitrogen and sulfur cycles all include at least one phase in which the element is in its gaseous state. Very little phosphorus circulates in the atmosphere because at Earth's normal temperatures and pressures, phosphorus and its various compounds are not gases. The largest reservoir of phosphorus is in sedimentary rock.

It is in these rocks where the phosphorus cycle begins. When it rains, phosphates are removed from the rocks (via weathering) and are distributed throughout both soils and water. Plants take up the phosphate ions from the soil. The phosphates then moves from plants to animals when herbivores eat plants and carnivores eat plants or herbivores. The phosphates absorbed by animal tissue through consumption eventually returns to the soil through the excretion of urine and feces, as well as from the final decomposition of plants and animals after death.

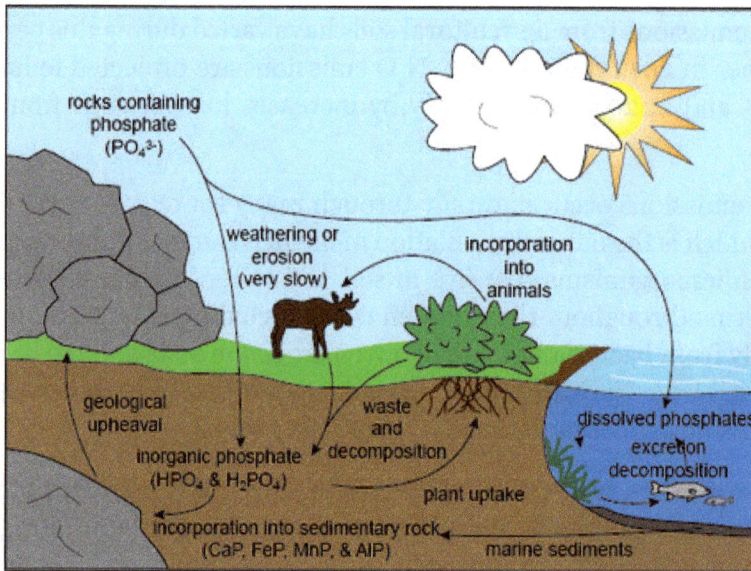

The same process occurs within the aquatic ecosystem. Phosphorus is not highly soluble, binding tightly to molecules in soil, therefore it mostly reaches waters by traveling with runoff soil particles. Phosphates also enter waterways through fertilizer runoff, sewage seepage, natural mineral deposits, and wastes from other industrial processes. These phosphates tend to settle on ocean floors and lake bottoms. As sediments are stirred up, phosphates may reenter the phosphorus cycle, but they are more commonly made available to aquatic organisms by being exposed through erosion. Water plants take up the waterborne phosphate which then travels up through successive stages of the aquatic food chain.

While obviously beneficial for many biological processes, in surface waters an excessive concentration of phosphorus is considered a pollutant. Phosphate stimulates the growth of plankton and plants, favoring weedy species over others. Excess growth of these plants tend to consume large amounts of dissolved oxygen, potentially suffocating fish and other marine animals, while also blocking available sunlight to bottom dwelling species. This is known as eutrophication.

Humans can alter the phosphorus cycle in many ways, including in the cutting of tropical rain forests and through the use of agricultural fertilizers. Rainforest ecosystems are supported primarily through the recycling of nutrients, with little or no nutrient reserves in their soils. As the forest is cut and/or burned, nutrients originally stored in plants and rocks are quickly washed away by heavy rains, causing the land to become unproductive. Agricultural runoff provides much of the phosphate found in waterways. Crops often cannot absorb all of the fertilizer in the soils, causing excess fertilizer runoff and increasing phosphate levels in rivers and other bodies of water. At one time the use of laundry detergents contributed to significant concentrations of phosphates in rivers, lakes, and streams, but most detergents no longer include phosphorus as an ingredient.

References

- Climate-systems-and-change, earthscience: lumenlearning.com, Retrieved 11 June, 2019

- Energyflows: nmt.edu, Retrieved 28 February, 2019

- Climate-Change, Impacts-on-Natural-Sulfur-Production, Ocean-Acidification-and-Community-Shifts: researchgate.net, Retrieved 3 April, 2019

- Nitrogen-cycle: noaa.gov, Retrieved 14 August, 2019

- Phosphorus-cycle, biogeochemical-cycles, air-climate-weather: enviroliteracy.org, Retrieved 7 March, 2019

Chapter 4

The Greenhouse Effect

The process through which radiation from a planet's atmosphere warms the planet's surface to a temperature above compared to what it would be otherwise is known as the greenhouse effect. This chapter closely examines the key concepts related to the greenhouse effect such as its causes and greenhouse gases to provide an extensive understanding of the subject.

Greenhouse Gases

Greenhouse gas is any gas that has the property of absorbing infrared radiation (net heat energy) emitted from Earth's surface and reradiating it back to Earth's surface, thus contributing to the greenhouse effect. Carbon dioxide, methane, and water vapour are the most important greenhouse gases. (To a lesser extent, surface-level ozone, nitrous oxides, and fluorinated gases also trap infrared radiation.) Greenhouse gases have a profound effect on the energy budget of the Earth system despite making up only a fraction of all atmospheric gases. Concentrations of greenhouse gases have varied substantially during Earth's history, and these variations have driven substantial climate changes at a wide range of timescales. In general, greenhouse gas concentrations have been particularly high during warm periods and low during cold periods.

A number of processes influence greenhouse gas concentrations. Some, such as tectonic activities, operate at timescales of millions of years, whereas others, such as vegetation, soil, wetland, and ocean sources and sinks, operate at timescales of hundreds to thousands of years. Human activities—especially fossil-fuel combustion since the Industrial Revolution—are responsible for steady increases in atmospheric concentrations of various greenhouse gases, especially carbon dioxide, methane, ozone, and chlorofluorocarbons (CFCs).

The effect of each greenhouse gas on Earth's climate depends on its chemical nature and its relative concentration in the atmosphere. Some gases have a high capacity for absorbing infrared radiation or occur in significant quantities, whereas others have considerably lower capacities for absorption or occur only in trace amounts. Radiative forcing, as defined by the Intergovernmental Panel on Climate Change (IPCC), is a measure of the influence a given greenhouse gas or other climatic factor (such as solar irradiance or albedo) has on the amount of radiant energy impinging upon Earth's surface.

Greenhouse Gas Intensity

Emission intensities vary widely across countries. Among the major emitters, GHG intensity varies almost seven-fold—from 344 tons per million dollars GDP in France, to 2,369 tons in Ukraine. France—with relatively low energy intensity, and very low carbon intensity, owing to its reliance on nuclear power—generates only 1.5 percent of global CO_2 emissions while producing 3.3 percent of global GDP. Ukraine—with high coal consumption and one of the world's most energy-intensive economies— generates 1.4 percent of global CO_2 emissions from only 0.5 percent of global GDP. As the country data suggests, however, intensity levels are unconnected with the size of a country's economy or population. A large or wealthy country may have a low GHG intensity, and vice-versa.

Like absolute and per capita emission levels, relative emission intensities vary depending on which gases are included. The inclusion of non-CO_2 gases boosts all countries' intensity levels, but in significantly different proportions. Aggregate CO_2 intensities are similar for developing and developed countries, while GHG intensities (using all six GHGs) in developing countries are about 40 percent higher, on average, than those in developed countries. Likewise, reported intensity levels depend on how GDP is measured. GDP may be expressed in a national currency, U.S. dollars, international dollars (using purchasing power parity conversions), or other common currency. Further, currencies may be inflation-adjusted to different base years. (To facilitate international comparisons, figures here use GDP measured in purchasing power parity expressed in constant 2000 international dollars.) Historically, emissions intensities fell between 1990 and 2002 for most countries, including three-fourths of the major emitters.

Among the top 25 emitters, carbon intensity dropped an average 15 percent, helping to drive a global decline of a commensurate amount. The most striking decline was in China, where intensity dropped 51 percent over the 12-year period. However, preliminary data for 2003 and 2004 shows that this trend is reversing, with emissions growing at twice the rate of economic output. Carbon intensity rose significantly from 1990 to 2002 in Saudi Arabia, Indonesia, Iran, and Brazil.

Table: Emissions Intensity Levels and Trends Top 25 GHG Emitting Countries.

GHG Intensity, 2000		% Change, 1990-2000	
Country	Tons of CO_2 eq. / $mil. GDP-PPP	Intensity (CO_2 only)	GDP
Ukraine	2,369	-6	-50
Russia	1,817	-5	-26
Iran	1,353	17	64
Saudi Arabia	1,309	45	32
Pakistan	1,074	4	55

China	1,023	-51	205
South Africa	1.0006	-3	27
Poland	991	-43	47
Australia	977	-16	51
Turkey	844	-2	42
Indonesia	799	22	62
Canada	793	-15	40
India	768	-9	87
South Korea	729	-2	100
United States	720	-17	42
Brazil	679	17	35
Argentina	659	-18	33
Mexico	586	-9	41
Spain	471	5	37
Germany	471	-29	22
EU-25	449	-23	27
United Kingdom	450	-29	32
Japan	400	-6	16
Italy	369	-10	20
France	344	-19	24
Developed	633	-23	29
Developing	888	-12	71
World	715	-15	36

GHG intensity includes emissions from six gases. GHG intensity and CO_2 intensity exclude CO_2 from international bunker fuels and land use change and forestry. GDP is measured in terms of purchasing power parity (constant 2000 international dollars).

Drivers of Emissions Intensity

Population and GDP as major determinants of a country's emissions and changes in its emissions over time. Emissions intensity—the level of greenhouse gas emissions per unit of economic output—is a composite indicator of two other major factors contributing to a country's emissions profile, namely energy intensity and fuel mix (Equation).

Equation:

$$\underset{\substack{\text{Carbon} \\ \text{Intensity}}}{\frac{CO_2}{GDP}} = \underset{\substack{\text{Energy} \\ \text{Intensity}}}{\frac{Energy}{GDP}} \times \underset{\text{Fuel Mix}}{\frac{CO_2}{Energy}}$$

Following on equation, CO_2 emissions intensity is a function of two variables. The first variable is energy intensity, or the amount of energy consumed per unit of GDP. This reflects both a country's level of energy efficiency and its overall economic structure, including the carbon content of goods imported and exported. An economy dominated by heavy industrial production, for instance, is more likely to have higher energy intensity than one where the service sector is dominant, even if the energy efficiencies within the two countries are identical. Likewise, a country that relies on trade to acquire (import) carbon-intensive goods will—when all other factors are equal—have a lower energy intensity than those countries that manufacture those same goods for export.

Energy-intensity levels are not well correlated with economic development levels. Transition economies, such as Russia and Ukraine, tend to have the highest energy (and carbon) intensities. Intensities in developing countries tend to be somewhat higher than in industrialized countries, owing largely to the fact that developing countries generally have a higher share of their GDP coming from energy-intensive manufacturing industries, such as basic metals. Industrialized countries, on the other hand, have greater shares of their economies comprised of lower-carbon service sectors.

The second component of emissions intensity is fuel mix or, more specifically, the carbon content of the energy consumed in a country. Coal has the highest carbon content, followed by oil and then natural gas. Accordingly, if two nations are identical in energy intensity, but one relies more heavily on coal than the other, its carbon intensity will be higher. Figure shows the breakdown of fuel mixes for selected countries. Countries vary widely in their use of fuels. Coal dominates in some countries (for example, China and South Africa), gas prevails in others (Russia), while other fuels—like hydropower, biomass, and other renewable sources presumed carbon-neutral—are significant in still other countries (Brazil, India). "Other renewable energy," which includes solar, wind, and geothermal, accounts for no more than 3.5 percent of total primary energy supply in any of the major emitting countries. Fuel mixes, it should be further noted, are highly correlated with countries' natural endowments of coal, oil, gas, and hydropower capacity, a topic addressed further.

Table highlights the relative contribution of energy intensity and fuel mix to overall carbon intensity changes. In the EU, declining carbon intensity reflects reductions in both energy intensity and carbon content (for example, the switch from coal to gas in the U.K.). In the United States, declines stem almost entirely from reduced energy intensity. In some cases, the two factors counterbalance one another. In India, for instance, the increased carbon content of fuels has nearly entirely offset the effect of reduced energy intensity. South Korea's case is virtually the opposite: the switch to lower carbon fuels has nearly offset a sizable increase in energy intensity. Globally, the decline in overall carbon intensity stems more from reduced energy intensity than from changes in fuel mix.

Using the decomposition analysis introduce, Figure shows in more detail the relative effects of energy intensity and fuel mix in shaping absolute emission trends. In several countries, it can be seen that declines in intensity were accompanied by significant increases in GDP, leading to increases in absolute CO_2 levels. The most notable case is China, where the effect of significant intensity declines was more than offset by substantial GDP growth. Likewise, the U.S. decline in carbon intensity (17 percent) was offset by increases in population and GDP.

When non-CO_2 gases are considered, additional factors beyond energy intensity and fuel mix affect emissions intensities and trends. For instance, CH_4 and N_2O emissions from agricultural sources might be influenced significantly by commodity prices and shifts in international livestock and grain markets. Land-use change and forestry emissions might also be influenced by domestic and international forces unrelated to the factors discussed above.

GDP Changes and Projections

Emissions intensities, at least with respect to energy and industrial emissions, are influenced primarily by shifts in energy intensity, economic structure, and fuel mix. It follows that emission intensities are not directly correlated with changes in activity levels (GDP and population). Even in the event of major GDP changes, changes in intensity levels may be modest. Absolute emission levels, on the other hand, are most strongly influenced by GDP shifts. When GDP rises, emissions also tend to rise correspondingly. This correlation is illustrated in figure for South Korea, where the effect of the 1998 Asian financial crisis is clearly apparent. GDP and CO_2 moved in tandem while carbon intensity was less affected. Because of this correlation, projections of carbon intensity tend to exhibit less uncertainty than absolute emission forecasts.

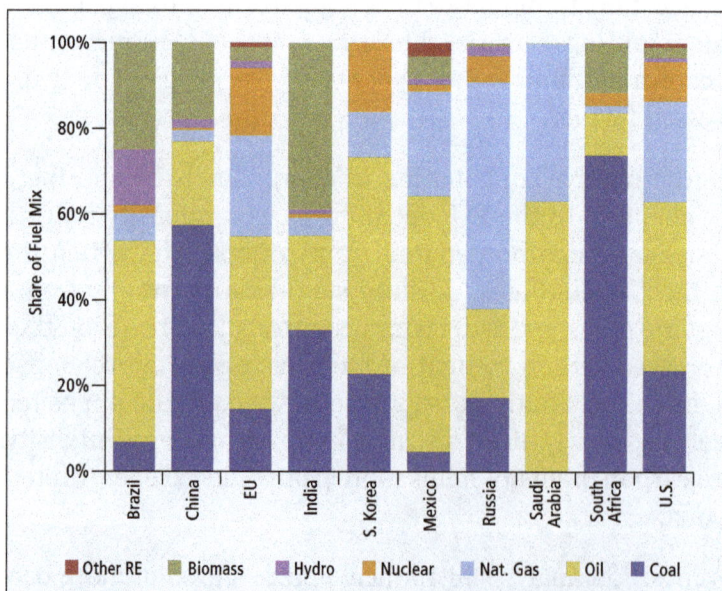

Fuel Mix in Energy Supply, selected major GHG emitters.

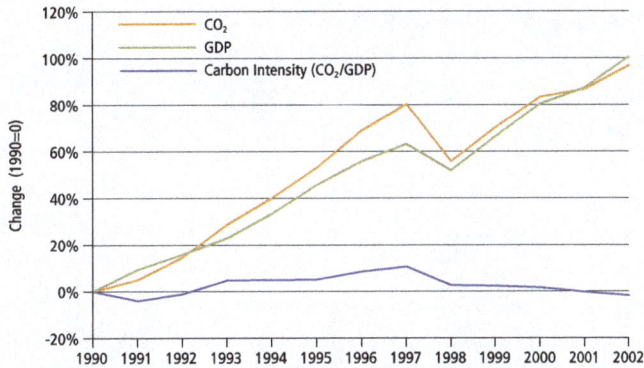

South Korea: Relationship between CO_2 and GDP.

This conclusion, however, may not hold in some instances. First, intensity projections may be less certain for countries whose national emissions profile includes large shares of non-CO_2 gases or LUCF-related emissions. As discussed above, these emissions are likely to be shaped by a different set of factors, many of which are difficult to predict. In general, non-CO_2 gases and LUCF-related emissions are not as strongly correlated with GDP.

Second, the uncertainty reduction benefits of intensity indicators may be less apparent for mature, developed economies. A simplified illustration can be made using projections from the United States EIA. Figure shows projections in terms of absolute emissions and emissions intensity for Japan, a mature economy, and nearby South Korea, a rapidly developing economy. For each country, projections include the EIA "high" and "low" growth scenarios. It is of course possible, if not likely, that all of the projections turn out to be significantly off the mark, as discussed. Nevertheless, the perceived "uncertainty" gap (i.e., difference between high and low) for absolute emissions is very large for South Korea, whereas the intensity gap is relatively small. In other words, according to the EIA, future emissions (in absolute terms) are highly uncertain in South Korea, whereas intensity is less so. For Japan, emissions are expected to grow between 5 and 18 percent by 2025. While not especially large, this uncertainty is not insignificant. What is notable, however, is that the uncertainty for Japan's intensity does not seem to be much less than for absolute emissions.

Implications for International Climate Cooperation

Emission targets, measured in intensity terms, can reduce cost uncertainty. Uncertainty is perhaps the most significant problem associated with target setting, and perhaps GHG mitigation in general. Not unjustifiably, governments tend to be unwilling to adopt commitments when it is unclear what kinds of policies and costs are implicit in those commitments. Framing a target in carbon intensity terms, as illustrated above, tends to reduce that uncertainty and, accordingly, may be a more attractive option than fixed targets. However, the reduced cost uncertainty comes at the expense of greater environmental uncertainty. Furthermore, the benefits of reduced uncertainty are likely to be much greater for developing countries than for developed countries, as discussed above.

For developing countries, a high proportion of emissions may come from non-CO_2 gases and landuse change and forestry. When these emissions are factored into intensity targets, the benefits of reduced uncertainty tend to be lower, since these emissions are less tied to economic activity levels than CO_2 from fossil fuels. The case of Argentina's proposed target illustrates this phenomenon. In 1999, Argentina sought to adopt a "dynamic" emission target under the Kyoto Protocol. However, CH_4 and N_2O from agriculture accounted for more than 40 percent of Argentina's GHG emissions. Future agricultural emissions would be influenced more by the international livestock and grain market than domestic GDP. Accordingly, Argentina chose not to propose a simple "intensity" target. Instead, Argentina suggested a complex indexing methodology tailored to their particular circumstances.

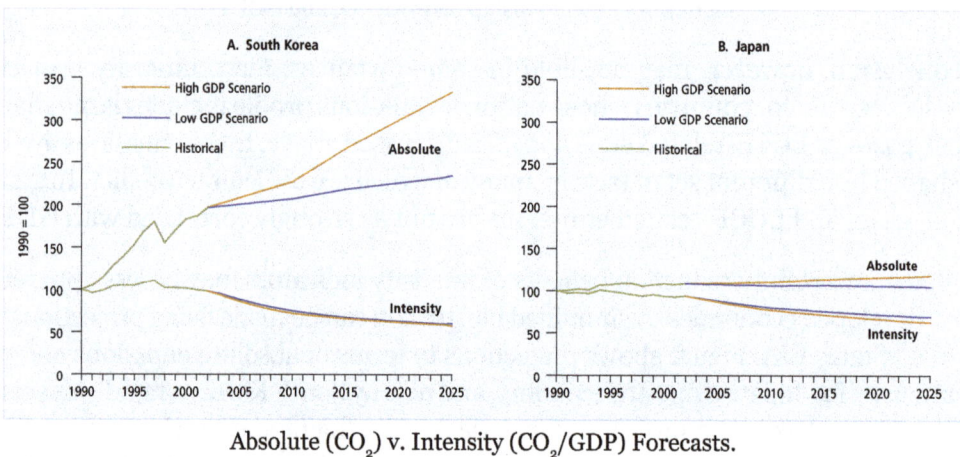

Absolute (CO_2) v. Intensity (CO_2/GDP) Forecasts.

GHG targets, measured in intensity terms, may introduce complexities and reduce transparency. Intensity targets would make international climate negotiations (and domestic policy-making) more complex, especially if they are being adopted by many countries. Countries might try to adopt both different percentage reduction commitments (as in Kyoto) and different GDP adjustment provisions, as the Argentine case illustrates. Negotiations might become exceedingly complex, to the point that non-specialists, or indeed anyone other than climate negotiators themselves, would have difficulty understanding proposed commitments.

The case of the Bush Administration's GHG intensity target helps illustrate the potential for confusion with this approach. The target—an 18 percent reduction in GHG intensity over the 2002 to 2012 timeframe—was introduced as a bold new effort. Yet, in the preceding 10-year period from 1992 to 2002, U.S. GHG emissions intensity had dropped by 18.4 percent and, assuming continued U.S. GDP growth, the target would permit U.S. emissions to rise by 14 percent over the decade. Thus, the Bush Administration's target is essentially a continuation of past trends; one that is likely to result in increases, not reductions, of GHG emissions in absolute terms. This has often been misunderstood or inaccurately reported in the U.S. media. Thus, more than some other metrics, intensity targets may be vulnerable to misperceptions and obfuscation.

Overall, intensity targets represent a trade-off in terms of benefits and drawbacks. In some instances intensity targets would significantly aid in uncertainty reduction, but at the expense of simplicity and transparency. Complexities of intensity targets also extend to other areas, not discussed above, such as interactions with international emissions trading.

Embedded Emissions

One way of attributing greenhouse gas (GHG) emissions is to measure the embedded emissions of goods that are being consumed (also referred to as "embodied emissions"). This is different from the question of to what extent the policies of one country to reduce emissions affect emissions in other countries (the "spillover effect" and "carbon leakage" of an emissions reduction policy). The UNFCCC measures emissions according to production, rather than consumption. Consequently, embedded emissions on imported goods are attributed to the exporting, rather than the importing, country. The question of whether to measure emissions on production instead of consumption is partly an issue of equity, i.e., who is responsible for emissions.

The 37 Kyoto Protocol Parties, listed in Annex B of the treaty, have agreed to legally binding emission reduction commitments. Under the UNFCCC accounting of emissions, their emission reduction commitments do not include emissions attributable to their imports., Wang and Watson asked the question, "who owns China's carbon emissions?" In their study, they suggested that nearly a quarter of China's CO_2 emissions might be a result of its production of goods for export, primarily to the USA but also to Europe. Based on this, they suggested that international negotiations based on within country emissions (i.e., emissions measured by production) may be "missing the point."

Recent research confirms that, in 2004, 23% of global emissions were embedded in goods traded internationally, mostly flowing from China and other developing countries to the U.S., Europe and Japan. Research by the Carbon Trust in 2011 revealed that approximately 25% of all CO_2 emissions from human activities 'flow' (i.e. are imported or exported) from one country to another. The flow of carbon was found to be roughly 50% emissions associated with trade in commodities such as steel, cement, and chemicals, and 50% in semi-finished/finished products such as motor vehicles, clothing or industrial machinery and equipment.

CO_2 Emissions from Various Fuels

Different fuels emit different amounts of carbon dioxide (CO_2) in relation to the energy they produce when burned. To analyze emissions across fuels, compare the amount of CO_2 emitted per unit of energy output or heat content.

Table: Pounds of CO_2 emitted per million British thermal units (Btu) of energy for various fuels.

Coal (anthracite)	228.6
Coal (bituminous)	205.7
Coal (lignite)	215.4
Coal (subbituminous)	214.3
Diesel fuel and heating oil	161.3
Gasoline (without ethanol)	157.2
Propane	139.0
Natural gas	117.0

The amount of CO_2 produced when a fuel is burned is a function of the carbon content of the fuel. The heat content, or the amount of energy produced when a fuel is burned, is mainly determined by the carbon (C) and hydrogen (H) content of the fuel. Heat is produced when C and H combine with oxygen (O) during combustion. Natural gas is primarily methane (CH_4), which has a higher energy content relative to other fuels, and thus, it has a relatively lower CO2-to-energy content. Water and various elements, such as sulfur and noncombustible elements in some fuels, reduce their heating values and increase their CO_2-to-heat contents.

Greenhouse Effect

Our Goldilocks Planet

Currently the earth has just the right amount of greenhouse gases ...

Venus:
Too much
Average surface temperature = 864 F

Earth:
Just right
57 F

Mars:
Too little
-80 F

Showing the temperature of some planets: Mars, Earth, and Venus.

The factor that Earth has an average surface temperature pleasurably between the boiling point and freezing point of water, therefore suitable for our kind of life, cannot be clarified by merely proposing that planet Earth orbits at just the precise space from the sun to absorb just the right amount of solar radiation. The moderate temperatures are also the outcome of having just the precise kind of atmosphere. The atmosphere in planet Venus would produce hellish, Venuslike conditions on planet Earth; the Mars troposphere would leave earth shivering in a Martian-type deep freeze.

Additionally, parts of the earth's atmosphere act as shielding blanket of just the right thickness, receiving appropriate solar energy to keep the global average temperature in an amusing range. The Martian blanket is too thin, and the Venusian blanket is way too thick. The 'blanket' as stated here, is termed as a collection of atmospheric gases called greenhouse gases based on the knowledge that the gases also capture heat similar to the glass walls of a greenhouse. These gases, mostly water vapor, carbon dioxide, methane, and nitrous oxide, all perform as effective global insulators.

The conversation of inbound and outward-bound radiation that warms the Earth is often referred to as the greenhouse effect because a greenhouse works in much the same way.

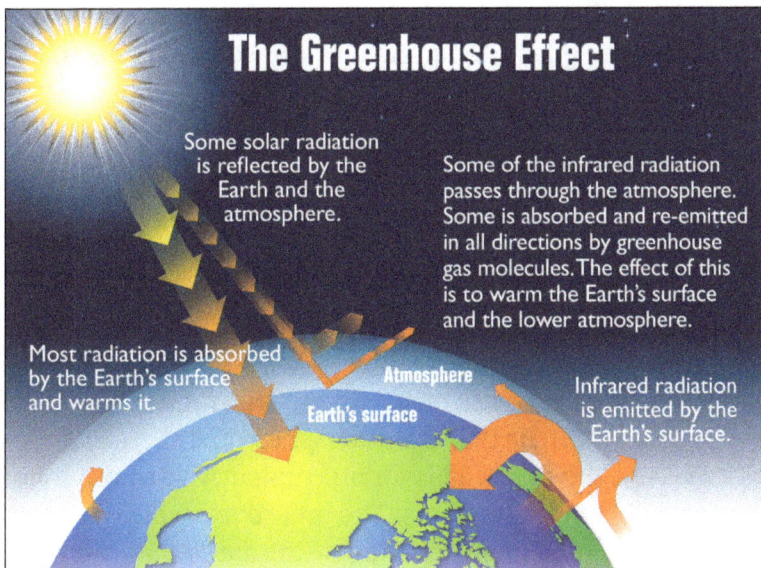

Showing radiation absorption and emission by greenhouse gases.

Inbound Ultra Violet (UV) radiation easily passes through the glass walls of a greenhouse and is absorbed by the plants and hard surfaces inside. Weaker Infrared (IR) radiation, however, has difficulty passing through the glass walls and is trapped inside, that is, warming the greenhouse. This outcome lets tropical plants flourish inside a greenhouse, even during a cold winter.

The greenhouse influence upsurges the temperature of the Earth by trapping heat in our atmosphere. This retains the temperature of the Earth higher than it would be if direct heating by the Sun was the only source of warming.

When sunlight reaches the surface of the Earth, some of it is absorbed which warms the ground and some jumps back to space as heat. Most Greenhouse gases that are in the atmosphere fascinate and then transmit some of this heat back towards the Earth.

The greenhouse effect is a foremost factor in keeping the Earth heartfelt because it keeps some of the planet's heat that would otherwise escape from the atmosphere out to space. In fact, without the greenhouse effect the Earth's average global temperature would be much colder and life on Earth as we recognize it would not be possible. The difference between the Earth's actual average temperature 14 °C (57.2 °F) and the expected effective temperature just with the Sun's radiation -19 °C (-2.2 °F) gives us the strength of the greenhouse effect, which is 33 °C.

The greenhouse effect is a natural process that is millions of years old. It plays a critical role in a variable the overall temperature of the Earth. The greenhouse effect was first discovered by Joseph Fourier in 1827, experimentally verified by John Tyndall in 1861, and quantified by Svante Arrhenius in 1896. It gives information about despite the looming difficult energy context in the majority of countries in the world, global change in environmental dignity resulting from power generation and energy consumption scenario is rapidly becoming a globally disturbing phenomenon. The present study focused on the greenhouse effect: the greenhouse gases and their impacts on global warming.

Foundations of Greenhouse Effect

The greenhouse effect is mostly caused by the interaction of the sun's energy with greenhouse gases such as carbon dioxide, methane, nitrous oxide and fluorinated gases in the Earth's atmosphere. The ability of these gases to capture heat is what causes the greenhouse effect.

Greenhouse gases consist of three or more atoms. This molecular structure makes it possible for these gases to trap heat in the atmosphere and then transfer it to the surface which further warms the Earth. This uninterrupted cycle of trapping heat clues to an overall increase in global temperatures. The procedure, which is very similar to the way a greenhouse works, is the main reason why the gases that can produce this outcome are collectively called as greenhouse gases.

The prime forcing gases of the greenhouse effect are: Carbon dioxide (CO_2), methane (CH_4), nitrous oxide (N_2O), and fluorinated gases.

Reaction Gas (Water Vapor) of the Greenhouse Effect

Carbon dioxide is to some extent one of the greenhouse gases. It involves one carbon atom with an oxygen atom bonded to each side. As soon as its atoms are bonded tightly together, the carbon dioxide molecule can absorb infrared radiation and the molecule starts to vibrate. Eventually, the vibrating molecule will emit the radiation again, and it will likely be absorbed by yet another greenhouse gas molecule. This absorption-emission-absorption cycle serves to keep the heat near the surface, effectively insulating the surface from the cold of space.

Carbon dioxide, water vapor (H_2O), methane (CH_4), nitrous oxide (N_2O), and some limited other gases are greenhouse gases. They all are molecules made up of more than two constituents atoms, bound loosely enough together to be able to vibrate with the absorption of heat. The foremost mechanisms of the atmosphere (N_2 and O_2) are two-atom molecules too closely bound together to vibrate and consequently, they do not absorb heat and subsidize to the greenhouse effect.

Carbon dioxide, methane, nitrous oxide and the fluorinated gases are all well-mixed gases in the atmosphere that do not react to changes in temperature and air pressure, so the levels of these gases are not affected by condensation effect. Water vapor also is a highly active component of the climate system that retorts briskly to fluctuations in conditions by either dwindling into rain or snow or evaporating to return to the atmosphere. Consequently, the imprint of the greenhouse effect is principally circulated through water vapor, and it turns as a fast reaction effect.

Carbon dioxide and the other non-condensing greenhouse gases are the vital gases within the Earth's atmosphere that tolerate the greenhouse effect and rheostat its strength. Water vapor is a fast-acting feedback but its atmospheric concentration is controlled by the radiative forcing supplied by the non-condensing greenhouse gases.

In fact, the greenhouse effect would collapse were it not for the presence of carbon dioxide and the other non-condensing greenhouse gases. Together the feedback by the condensing and the forcing by the non-condensing gases within the atmosphere both play an important role in the greenhouse effect.

Reduction of Greenhouse Gases

The primary objective of WWTPs is to meet effluent standards. In order to protect the receiving water body. However, reduction of GHG emissions from WWTPs requires a broadening in scope. The estimated quantity of N_2O from WWTPs by the United States Environmental Protection Agency. Accounts for approximately 3% of N_2O from all national sources which rank as the sixth largest contributor to GHG emissions. The right quantification of GHG is a necessity to better understand how to effectively reduce GHG emissions from WWTPs, as well as to improve the accuracy in the GHG emission reporting processes. There is keen interest in climate change issues due to a fast increasing rate of GHG emissions. This has emphasized the need to innovate and establish right approaches to better design, control and optimize WWTPs on the plant-wide scale.

In recent years, one of the cheap modern and promising solutions to decreasing GHG emission into the Earth's atmosphere is the employment bioremediation technique. Other mitigation plans to avert the negative outcomes of greenhouse effect may include activities such increase in tree planting, reduction in burning fossil fuels, exploitation of affordable, clean and renewable of energy, carbon dioxide capture and sequestration etc.

Bioremediation technique employs microbial metabolism to remove pollutants. A bioremediation technique and strategy (phytoremediation enhanced by endophytic microorganisms) can be used to remove hazardous waste including greenhouse gases from the biosphere. Phytoremediation is the most effective bioremediation technique employed to remove greenhouse gases. In phytoremediation, living green plants in situ are used. Living green plants have the ability to decrease or remove contaminants from soil, air, water, and sediments. Recently, selected or engineered endophytic microorganisms have been used to improve the phytoremediation processes. Many studies have demonstrated the efficacy of endophytic microorganisms in accelerating these processes by interacting closely with their host plants.

Another technique for reducing the negative effects of the greenhouse effect is to use methanotrophic endophytes inhabiting Sphagnum Spp. which can act as a natural methane filter. It can reduce CH_4 and CO_2 emission from peatlands by up to 50%. Studies have demonstrated potential ability of the plant–methanotrophic bacteria systems in the reduction of methane emission up to 77%, depending on the season and the host plant.

Current Existing Challenges to Reducing Greenhouse Gases (GHG)

Currently, there are difficulty challenges in controlling GHG emissions for different WWTPs. Measurement uncertainties and lack of transposable data still hinder a correct and required GHG emission quantification process.

One recommendation to fill this gap includes the use of mathematical models which offer useful tools for assessing GHG and evaluating different mitigation alternatives before putting them into practice. GHG modelling can enhance the correct quantification of GHG emissions for different WWTP configurations and evaluate the effects of various operating conditions. In recent years, a large portfolio of mathematical modelling studies has been developed to include GHG emissions during design, operation, and optimization of WWTPs.

Admonished the scientific community to examine the key elements of GHG modelling using a plant-wide approach. Several advantages and potentials of this approach include: i) An approach which takes into account the role of each plant treatment unit process and the interactions among them and ii) Operation or control of each particular unit, not only at local level but as a component of a system, and avoids the risk of a sub-optimization (an example is a reduction of effluent quality at higher operational costs.

Solar Radiation

The sun radiates gigantic quantities of energy into space, crosswise a wide spectrum of wavelengths.

Utmost of the radiant energy from the sun is concentrated in the visible and near-visible portions of the spectrum. The narrow band of visible light, between 400 and 700 nm, signifies 43% of the total radiant energy emitted. Wavelengths shorter than the visible account for 7 to 8% of the total, but are extremely important because of their high energy per photon. The shorter the wavelength of light, the more energy it contains. Accordingly, ultraviolet light is very energetic (accomplished by breaking apart stable biological molecules and instigating sunburn and skin cancers). The residual 49 - 50% of the radiant energy is spread over the wavelengths longer than those of visible light. These lie in the near infrared range from 700 to 1000 nm; the thermal infrared, between 5 and 20 microns; and the far infrared regions. Various components of earth's atmosphere absorb ultraviolet and infrared solar radiation before it penetrates to the surface, but the atmosphere is quite transparent to visible light.

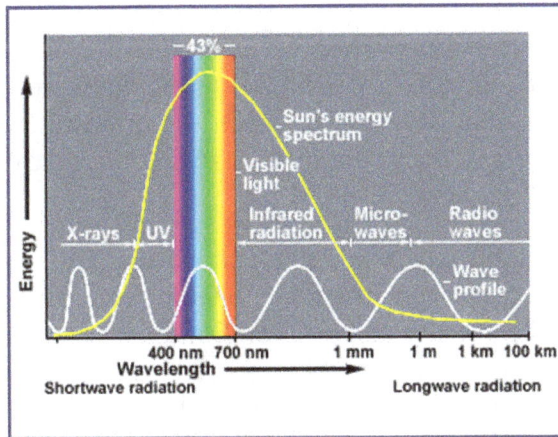

Showing the wave profile of various radiations.

Absorbed by land, oceans, and vegetation at the surface, the visible light is transformed into heat and re-radiates in the form of invisible infrared radiation. During the day, earth heats up, but at night, all the accumulated energy would radiate back into space and the planet's surface temperature would fall far below zero very rapidly. The reason this doesn't happen is that earth's atmosphere contains molecules that absorb the heat and re-radiate the heat in all directions. This reduces the heat radiated out to space called greenhouse gases because they serve to hold heat in like the glass walls of a greenhouse, these molecules are responsible for the fact that the earth enjoys temperatures suitable for our active and complex biosphere.

Sources of Greenhouse Gas Emissions

In recent times, one of the major sources of greenhouse gas (GHG) emission is from water resource recovery facilities (wastewater treatment plants (WWTPs). Wastewater treatment plants (WWTPs) are recognized as one of the larger minor sources of GHG emissions. The WWTPs emit gases such as nitrous oxide (N_2O), carbon dioxide (CO_2), and methane (CH_4). Increasing emission of GHG from this source poss harm to our climate.

Biological mechanisms such as emissions of CO_2 due to microbial respiration, emission of N_2O by nitrification and denitrification, and emission of CH_4 from anaerobic digestion processes are direct emissions from WWTPs. Sources that not regulated directly within the WWTP are indirect internal emission sources; consumption of thermal energy and indirect external emission sources; third-party biosolids hauling, chemical productions and their transportation to the plant, etc.

The increasing rate of GHG emissions is due to the changes in the economic output, extended energy consumption, increasing emission from landfills, livestock, rice farming, septic processes, and fertilizers as well as other factors. Increase industrialisation, use of fertilizers, burning of fossil fuels and other human and natural activities result in a rise above normal average atmospheric temperature; thus posing threat to our environment.

Research identifies methane and carbon dioxide as the main greenhouse gases. Therefore, the reduction of methane concentration in the atmosphere, both from natural and anthropogenic sources, is indispensable to tackle the negative outcomes of global warming.

Atmospheric scientists first used the word 'greenhouse effect' in the later 1800s. At that time, it was used to designate the naturally happening functions of trace gases in the atmosphere and did not have any negative implications. It was not up until the mid-1950s that the term greenhouse effect was attached to concern over climate alteration. And in contemporary decades, we often hear about the greenhouse effect in somewhat negative terms. The negative concerns are related to the possible impacts of an improved greenhouse effect. It is important to remember that without the greenhouse effect, lifecycle on earth as we know it would not be possible.

While the earth's temperature is reliant on upon the greenhouse-like action of the atmosphere, the extent of heating and cooling are toughly influenced by several factors just as greenhouses are pretentious by various factors.

In the atmospheric greenhouse effect, the type of surface that sunlight first happenstances are the most important factor. Forests, grasslands, ocean surfaces, ice caps, deserts, and cities all absorb, reflect, and radiate radiation differently. Sunlight falling on a white glacier surface strongly reflects back into space, resulting in minimal heating of the surface and lower atmosphere. Sunlight falling on a dark desert soil is strongly absorbed, on the other hand, and contributes to significant heating of the surface and lower atmosphere. Cloud cover also affects greenhouse warming by both reducing the amount of solar radiation reaching the earth's surface and by reducing the amount of radiation energy emitted into space.

Scientists outline the percentage of solar energy reflected back by a surface. Understanding local, regional, and global effects are life-threatening to foretelling global climate change.

Greenhouse Gases and Global Warming

Greenhouse gases (GHGs) such as carbon dioxide, methane, nitrous oxide, and halogenated compounds emissions are caused by human activities and some do occur naturally. The GHGs absorb infrared radiation and trap heat in the atmosphere, thereby enhancing the natural greenhouse effect defined as global warming. This natural occurrence warms the atmosphere and make life on earth possible, without which the low temperature will make life impossible to live on earth.

"Gas molecules that captivate thermal infrared radiation, and are in a substantial amount, can force the climate system. These type of gas molecules are called greenhouse gases," Michael Daley, an associate professor of Environmental Science at Lasell College told Live Science. Carbon dioxide (CO_2) and other greenhouse gases turn like a blanket, gripping Infrared (IR) radiation and preventing it from evading into outer space. The net effect is the steady heating of Earth's atmosphere and surface, and this process is called global warming.

These greenhouse gases include water vapor, CO_2, methane, nitrous oxide (N_2O) and other gases. Since the dawn of the Industrial Revolution in the early 1800s, the scorching of fossil fuels like coal, oil, and gasoline have greatly increased the concentration of greenhouse gases in the atmosphere, specifically CO_2, National Oceanic and Atmospheric Administration (NOAA). "Deforestation is the second largest anthropogenic basis of carbon dioxide to the atmosphere ranging between 6% and 17%," said Daley.

Some human activities like the production and consumption of fossil fuels, use of various chemicals agriculture, burning bush, waste from incineration processes and other industrial activities have increased the concentration of greenhouse gases (GHG), particularly CO_2, CH_4, and N_2O in the atmosphere making them harmful.

This increase in atmospheric GHG concentration has led to climate change and global warming effect, which is motivating international efforts such as the Kyoto Protocol, signing of Paris Agreement on climate change and other initiatives to control negative outcomes of the greenhouse effect. The contribution of a greenhouse gas to global warming is commonly expressed by its global warming potential (GWP) which enables the comparison of global warming impact of the gas and that of a reference gas, typically carbon dioxide.

Atmospheric CO_2 intensities have increased by more than 40% since the beginning of the Industrial Revolution, from about 280 parts per million (ppm) in the 1800s to 400 ppm today. The last time Earth's atmospheric levels of CO_2 reached 400 ppm was during the Pliocene Epoch, between 5 million and 3 million years ago, according to the University of California, San Diego's Scripps Institutions of Oceanography.

The greenhouse effect, collective with growing levels of greenhouse gases and the resultant global warming, is expected to have profound consequences, according to the near-universal consensus of scientists.

If global warming undergoes unimpeded, it will cause noteworthy climate change, a rise in sea levels, increasing ocean acidification, lifethreatening weather events and other severe natural and societal impacts, according to NASA, the Environmental Protection Agency(EPA) and other scientific and governmental bodies.

Can the Greenhouse effect be Overturned?

Several scientists approve that the impairment of the Earth's atmosphere and climate is long-gone the point of no reoccurrence or that the destruction is near the point of no return. "I agree that we have passed the point of avoiding climate change," Josef Werne, an associate professor at the department of geology and planetary science at the University of Pittsburgh. In Werne's opinion, there are three options from this point forward:

1. Do nothing and live with the moments.

2. Acclimatize to the changing climate (which includes things like rising sea level and related flooding).

3. Alleviate the impact of climate change by belligerently enacting policies that actually reduce the concentration of CO_2 in the atmosphere.

Keith Peterman, a professor of chemistry at York College of Pennsylvania, and Gregory Foy, an associate professor of chemistry at York College of Pennsylvania believes that the damage isn't to that point yet and that international agreements and action can save the planet's atmosphere.

Causes of Greenhouse Effect

1. Burning of Fossil Fuels: Fossil fuels like coal, oil and natural gas have become an integral part of our life. They are used on large basis to produce electricity and for transportation. When they are burnt, the carbon stored inside them is released which combines with oxygen in the air to create carbon dioxide. With the increase in the population, the number of vehicles have also increased and this has resulted in increase in the pollution in the atmosphere. When these vehicles run, they release carbon dioxide, which is one the main gas responsible for increase in greenhouse effect.

    Apart from that, electricity-related emissions are high because we are still dependent on coal for electricity generation which releases large amount of CO_2 into the atmosphere and is still the primary source of fuel for generating electricity. Although, renewable sources are catching up, but it may take a while before we can reduce our dependance on coal for electricity generation.

2. Deforestation: Forests hold a major green area on the planet Earth. Plants and trees intake carbon dioxide and release oxygen, through the process of photosynthesis, which is required by humans and animals to survive. Large scale development has resulted in cutting down of trees and forests which has forced people to look for alternate places for living. When the wood is burnt, the stored carbon in converted back into carbon dioxide.

3. Increase in Population: Over the last few decades, there have been huge increase in the population. Now, this has resulted in increased demand for food, cloth and shelter. New manufacturing hubs have come up cities and towns that release some harmful gases into the atmosphere which increases the greenhouse effect. Also, more people means more usage of fossil fuels which in turn has aggravated the problem.

4. Farming: Nitrous oxide is one the greenhouse gas that is used in fertilizer and contributes to greenhouse effect which in turn leads to global warming.

5. Industrial Waste and Landfills: Industries which are involved in cement production, fertilizers, coal mining activities, oil extraction produce harmful greenhouse gases. Also, landfills filled with garbage produce carbon dioxide and methane gas contributing significantly to greenhouse effect.

Increased greenhouse gases directly increase the heat on the planet's surface and lower atmosphere. This has a rippling effect as it can thin and even create holes in the ozone layer. This means that other radiation like ultra violet (UV) rays can seep in from the Sun.

What this eventually means for life on Earth is that it will have to adapt to increasing temperatures. We already know that life is very adaptive, but we don't know for how long the ozone will continue to be depleted or at what rate. More heat means more fossil fuels will be burnt to cool down the heat. The burning of these fossil fuels will again produce more greenhouse gases and affect the environment adversely.

Many scientists believe that global warming and increased greenhouse effect are simply part of a global cycle. It seems like Earth goes through these cycles every so often of warming and then cooling. It is very difficult to determine what effect man-made gases have on the ozone layer and the greenhouse effect when so much of naturally occurring systems cause numerous effects themselves.

If the Earth does not go into a cooling cycle within the next few hundred years, it is possible that life on Earth might be very difficult for the generations to come. Increase in heat and radiation may make going outside difficult or dangerous during the daytime. These increased energies could also affect entire ecosystem. If plants and animals cannot adapt quickly enough.

Radiative Forcing

Radiative forcing is a measure as defined by the Intergovernmental Panel on Climate Change (IPCC), of the influence a given climatic factor has on the amount of downward-directed radiant energy impinging upon Earth's surface. Climatic factors are divided between those caused primarily by human activity (such as greenhouse gas emissions and aerosol emissions) and those caused by natural forces (such as solar irradiance). For each factor, so-called forcing values are calculated for the time period between 1750 and the present day. "Positive forcing" is exerted by climatic factors that contribute to the warming of Earth's surface, whereas "negative forcing" is exerted by factors that cool Earth's surface.

On average about 342 watts of solar radiation strike each square metre of Earth's surface per year, and this quantity can in turn be related to a rise or fall in Earth's surface temperature. Temperatures at the surface may also rise or fall through a change in the distribution of terrestrial radiation (that is, radiation emitted by Earth) within the atmosphere. In some cases, radiative forcing has a natural origin, such as during explosive eruptions from volcanoes where vented gases and ash block some portion of solar radiation from the surface. In other cases, radiative forcing has an anthropogenic, or exclusively human, origin. For example, anthropogenic increases in carbon dioxide, methane, and nitrous oxide are estimated to account for 2.3 watts per square metre of positive radiative forcing. When all values of positive and negative radiative forcing are taken together and all interactions between climatic factors are accounted for, the total net increase in surface radiation due to human activities since the beginning of the Industrial Revolution is 1.6 watts per square metre.

Radiative Damping

The sensitivity of the climate is essentially controlled by the so-called feedback parameter, which is the rate of radiative damping of the unit anomaly of the global mean surface temperature due to the outgoing radiation from the top of the atmosphere (TOA). By dividing the radiative forcing of climate by the feedback parameter, one gets the radiatively forced, equilibrium response of global surface temperature. This implies that the stronger is the rate of the radiative damping, the smaller is its equilibrium response to a given radiative forcing.

In the absence of feedback effect, the outgoing radiation at the top of the atmosphere is approximately equal to the fourth power of the effective planetary emission temperature, following the Stefan-Boltzmann's law of blackbody radiation. In the actual atmosphere, however, it deviates significantly from the blackbody radiation. When the temperature of the atmosphere increases, for example, its absolute humidity is likely to increases. Thus, the infrared opacity of the atmosphere increases, thereby lowering the temperature of the effective source of outgoing radiation and weakening the radiative damping of the surface temperature anomaly. This explains why the water vapor feedback weakens the radiative damping of surface temperature anomaly, thereby enhancing the sensitivity of climate.

The changes in the temperatures of the atmosphere and the earth's surface affect not only the outgoing longwave radiation but also the reflected solar radiation at the TOA. For example, an increase in surface temperature is likely to reduce the area covered by snow and sea ice, thereby reducing the heat loss due to the reflection of incoming solar radiation. Thus, the effective radiative damping of surface temperature anomaly is reduced, thereby enhancing the sensitivity of climate.

According to the third IPCC report, the previously estimated range of the equilibrium response of the global mean surface temperature to the doubling of atmospheric CO_2 has not reduced substantially over the last decade and remains between 1.5C and 4.5C. Clearly, the large range in the estimated sensitivity of surface temperature is attributable in no small part to our inability to reliably determine the influence of feedback upon the radiative damping of surface temperature anomaly.

Using the TOA fluxes of radiation obtained from the Earth radiation budget experiment (ERBE), the present study evaluates how the overall feedback of the atmosphere alters the radiative damping of the annual variation in surface temperature. Specifically, we compute the gain factor, which indicates the relative contribution of the overall feedback for reducing the radiative damping of the annual variation in global surface temperature. To identify the systematic bias of the overall feedback simulated by a model, the gain factor thus estimated is then compared with the gain factor of the feedback simulated by the model.

It is well-known that the annual variation of surface temperature is highly transient response to annually varying insolation that is out of phase between the two hemispheres. Thus, it is not our intension to determine the magnitude of feedback, incorrectly assuming that surface temperature were continuously in equilibrium with the annually varying, incoming solar radiation. Instead, we estimate here the magnitude of the overall feedback that operates upon the annual temperature variation, using the outgoing fluxes of terrestrial and reflected solar radiation from the TOA.

The annual variation of the global mean surface temperature is attributable mainly to the difference in effective thermal inertia between the two hemisphere than to the small annual variation of globally averaged, incoming solar radiation. Because the seasonal variation of surface temperature is much larger over continents than over oceans, the annual variation of the global mean surface temperature is dominated by the contribution from the continents in Northern Hemisphere. Its annual range is about 3.3 °C with highest temperature in July and the lowest in January. The range is comparable in magnitude to a current estimate of the equilibrium response of global mean surface temperature to the doubling of CO_2 concentration in the atmosphere.

Since the pattern of the annual variation of surface temperature differs greatly from that of the global warming simulated by a model, it is quite likely that the rate of the radiative damping of the global mean surface temperature anomaly is significantly different between the two. As noted by Raval and Ramanathan and Inamdar and Ramanathan, the rate of radiative damping of local surface temperature anomaly is similar to the damping of the annual variation in global surface temperature under clear sky. Therefore, it is likely that the rate of the radiative damping of global surface temperature variation under clear sky is similar between the annual variation and global warming despite the difference in pattern. On the other hand, a similar statement may not be made for the albedo-, and cloud feedback. Nevertheless, we are going to estimate the gain factor of the overall feedback for global warming using the gain factor for the annual variation, which is the largest climate change one can observe. The availability of data from the ERBE is another decisive factor for conducting the analysis presented here.

The gain factor of the overall feedback thus obtained is then compared with the gain factor of feedback simulated by a climate model. Similar comparison may be made for other feedback processes such as albedo- water vapor-, and cloud-feedbacks. We hope that such comparison should be useful for identifying the systematic bias of a model, thereby serving as a guide for improving the parameterization of the feedbacks that control the sensitivity of climate.

Formulation of Feedback Parameter

The sensitivity of climate may be determined by the feedback parameter, which is defined as the rate of radiative damping of global surface temperature anomaly at the TOA. From the given feedback parameter, one can compute the equilibrium response of global mean surface temperature, dividing the radiative forcing of climate by the feedback parameter.

$$\lambda = \frac{d\left(\overline{L} + \overline{[S_r]}^A\right)}{d\overline{T}\,\mathrm{s}},$$

Where L and $[S_r]^A$ denote the outgoing flux of longwave radiation and annually normalized flux of reflected solar radiation at the TOA (S_r), respectively. T_S denotes surface temperature. $\overline{(\ )}$ indicates the global average operator. Following Cess et al., the annual normalization is defined as follows:

$$[S_r]^A = \frac{(S_i)^A}{S_i} \times S_r,$$

Where S_i is the TOA flux of incoming solar radiation, and $[\ ]^A$ and $(\ )^A$ indicate the annual normalization and annual averaging, respectively.

The annual variation of the TOA flux of reflected solar radiation (S_r) is attributable not only to the annual variation in the state of the atmosphere-surface system but also to that of the incoming solar radiation (S_i). To extract the contribution to the annual variation from the former without the latter, it is necessary to annually normalize the reflected flux of solar radiation as indicated by Eq. 2, removing the direct contribution from the annual variation of the incoming solar radiation. (One should note here that the seasonal variation of the sun's zenith angle affects not only the incoming flux of solar radiation at the top of the atmosphere but also the albedos of the earth's surface and cloud cover. In addition to removing the former effect as we did, it is desirable to remove the latter effect. We did not do so because of the difficulty involved.)

The TOA flux of outgoing longwave radiation $\overline{L}$ may be subdivided into two components as follows:

$$\overline{L} = \varepsilon\sigma\overline{T}_s^4 + L^{FB},$$

where the first term represents the black body emission of the planet, and the second term (L^{FB}) denotes the contribution from feedback. e is the coefficient of planetary emission and is chosen such that the first term on the right hand side of the $\overline{L} = \varepsilon\sigma\overline{T}_s^4 + L^{FB}$, is equal to the TOA flux of outgoing longwave radiation, given the realistic distribution of temperature in the atmosphere.

The feedback parameter (λ) may be subdivided as follows:

$$\lambda = \lambda_L + \lambda_S.$$

where,

$$\lambda_L = 4\varepsilon\sigma\overline{T}_s^3 + \frac{dL^{FB}}{d\overline{T}s}$$

$$\lambda_S = \frac{d\overline{[S_r]}^A}{d\overline{T}s}$$

Following Hansen et al., the feedback parameter may be related to the gain factor (f) that represents the influence of feedback upon the radiative damping of global mean surface temperature anomaly.

$$\lambda = \lambda_0 (1 - f),$$

Where,

$$\lambda_0 = 4\varepsilon\sigma\overline{T}_s^3$$

As noted above, the feedback parameter (λ) is inversely proportional to the equilibrium response of the global mean surface temperature to a radiative forcing (i.e., the sensitivity of climate). Based upon $\lambda = \lambda_0 (1 - f)$, feedback is positive and enhances the sensitivity of climate, if gain factor (f) is positive. On the other hand, it is negative, if gain factor is negative.

Referring to Eqs. $\lambda = \lambda_L + \lambda_S., \lambda_L = 4\varepsilon\sigma\overline{T}_s^3 + \frac{dL^{FB}}{d\overline{T}s}, \lambda_S = \frac{d\overline{[S_r]}^A}{d\overline{T}s}$

and $\lambda = \lambda_0 (1 - f)$, the gain factor may be subdivided into longwave and solar gain factors (i.e., f_L and f_S) as follows:

$$f = f_L + f_S,$$

Where,

$$f_L = -\frac{1}{\lambda_0} \frac{dL^{FB}}{d\overline{T}s}$$

$$f_S = -\frac{1}{\lambda_0} \frac{d\overline{[S_r]}^A}{d\overline{T}s}$$

To represent the contribution from individual feedback, the gain factor (f) may be subdivided further as follows.

$$f = f_{LR} + f_{WV} + f_a + f_C,$$

where f_{LR}, fWV, f_a, and f_C represents the contribution from the lapse rate-, water vapor-, albedo-, and cloudfeedback, respectively. The longwave and solar components of cloud gain factors may be represented by the derivatives of longwave and solar cloud forcings with respect to the global mean surface temperature, respectively.

The monthly mean, TOA fluxes of solar and longwave radiation are computed for each month of the year at each grid point, using the data obtained from the ERBE mounted on ERB and NOAA satellite over the period from February 1985 to February 1990. The monthly mean global mean surface temperature is computed for each month of the year based upon the reanalysis of past daily weather data over the period from January 1982 to December 1994. The reanalysis was recently conducted jointly by the National Center for Environmental Prediction and National Center for Atmospheric Research.

Using the Equation $\lambda_s = \dfrac{d\overline{[S_r]^A}}{d\overline{T}s}$ and $f_s = -\dfrac{1}{\lambda_0}\dfrac{d\overline{[S_r]^A}}{d\overline{T}s}$

λ_s and f_s are computed from the slope of the regression between annually normalized, global mean reflected solar radiation and the global mean surface temperature.

Using the equation $\lambda_L = 4\varepsilon\sigma\overline{T}_s^3 + \dfrac{dL^{FB}}{d\overline{T}s}$ and $f_L = -\dfrac{1}{\lambda_0}\dfrac{dL^{FB}}{d\overline{T}s}$,

λ_s and f_L are computed from the slope of the regression between the globally averaged, outgoing flux of terrestrial radiation and the global mean surface temperature. To compute the annually normalized, reflected solar radiation from equation,

$$[S_r]^A = \frac{(S_i)^A}{S_i} \times S_r,$$

it is necessary to know the planetary albedo (reflectivity of solar radiation at the TOA) throughout the year. Obviously, it is impossible to determine the planetary albedo during a polar night in high latitudes. In the present study, we assumed that planetary albedo remains unchanged during a polar night after it reaches the peak value in the fall.

Using the data from the ERBE, the globally averaged monthly mean flux of outgoing terrestrial radiation is computed for all 12 months of the year, and is plotted against the global mean surface temperature in figure. This figure shows that, over the global scale, the outgoing radiation at the TOA increases with increase in surface temperature. The slope (with its standard error) of the regression line through the plots is 2.1±0.17 W m^{-2} K^{-1}, and this is substantially less than the slope for the blackbody radiation, which is 3.3 W m^{-2} K^{-1}. Given this slope, one can compute the longwave gain factor (f_L) (with its standard error) as 0.38±0.05. This result implies that the atmosphere affects outgoing terrestrial radiation in such a way that it enhances the annual variation of global mean surface temperature.

Globally averaged monthly mean fluxes of annually normalized, reflected solar radiation [Sr] ^{A} are computed from the ERBE data for 12 months of the year and are plotted in figure below against the monthly mean, global mean surface temperature. This figure shows that annually normalized, reflected solar radiation decreases with increasing global mean surface temperature. This result implies that reflection of solar radiation also acts in such a way that it enhances the annual variation of global mean surface temperature. It is likely that this positive feedback effect is attributable to the albedo feedback effect of snow and sea ice, which reflects a large fraction of solar radiation. The slope (with standard error) of the regression line through the plotted points in Fig. 1b is -1.07±0.07 W m^{-2} K^{-1}, implying that the solar gain factor (f_s) is 0.32±0.02.

In figure below, the monthly mean value of total outgoing radiation $\overline{L} + \left[S_r \right]^A A$ is globally averaged, and is plotted against global mean surface temperature. The slope of a regression line through the plotted points is 0.98±0.20 W m^{-2} K^{-1}, yielding the total gain factor (f) of 0.7±0.06. Summing up the solar and longwave gain factors obtained, one also gets a total gain factor of 0.7. The result presented here indicates that both solar and longwave feedbacks act in such a way as to enhance the annual variation of the global mean surface temperature. Thus, the combined solar and longwave damping of the annual variation of the global mean surface temperature anomaly is only 30% of the damping by blackbody radiation.

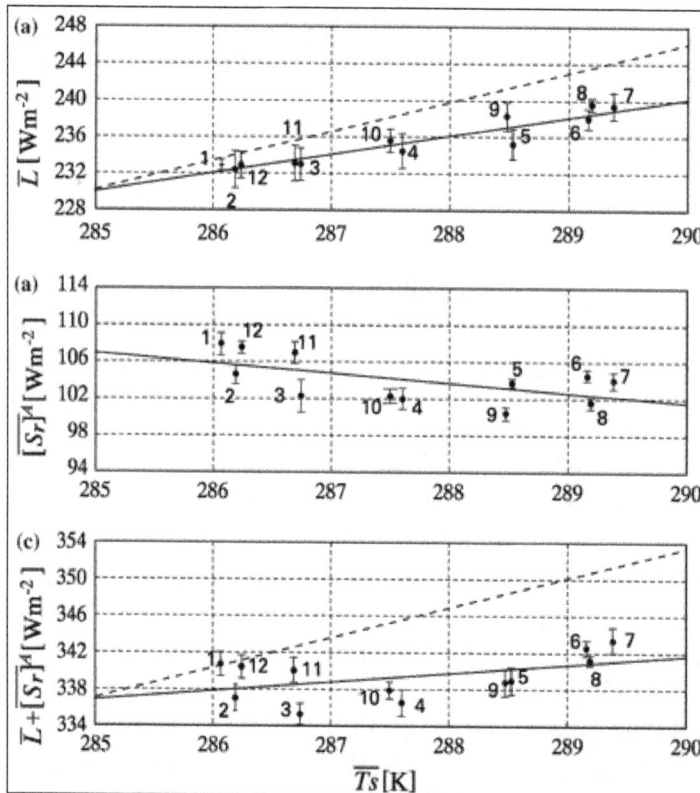

In the study on the cloud feedback, we used the data from the three models among many general circulation models of the atmosphere submitted to the Atmospheric Model Inter-comparison Project (AMIP)-I. They are CCSR 5.4.02 of the Center for Climate System Research/National Institute for Environmental Studies (CCSR/NIES), MPI-ECHAM 3 of the Max Planck Institute for Meteorology (MPI), and HAD-AM 1 of the United Kingdom Meteorological Office (UKMO). These models are chosen for the cloud feedback study because they explicitly predict the microphysical properties of cloud. The outputs from the time integration of these three models (with prescribed, seasonally varying sea surface temperature) are used again in the present study. It is not our intension to conduct here the comprehensive analysis of the overall feedback obtained from many models submitted to AMIP.

Using a bar diagram, figure above illustrates the gain factors, which are obtained from both GCMs and ERBE observation. It shows that the gain factors of overall feedback effect (f) from the three GCMs are approximately similar to the value from ERBE observation. However, when the gain factor is subdivided into solar and terrestrial components (i.e., f_s and f_l), the results are quite different from the observation. While solar and terrestrial gain factors obtained from the ERBE observations are similar to each other, the solar gain factors of all three models are smaller than the terrestrial gain factors. In the MPI model, for example, the solar gain factor is -0.08, quite different from the terrestrial gain factor, which is 0.66.

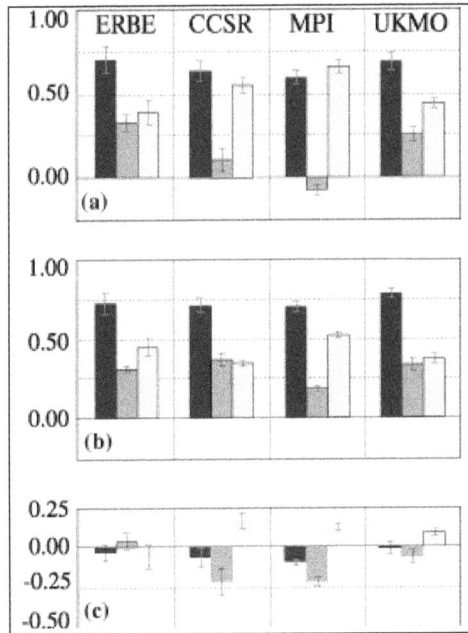

Tsushima and Manabe computed the cloud gain factors from the regression slopes between cloud radiative forcing and surface temperature over the domain between 60 °N and 60 °S. In the present study, we have repeated this computation, extending the domain to the entire globe. The result from the new analysis is illustrated. Despite the expansion of the analysis domain, the cloud gain factors computed from the ERBE data remain small, and are hardly different in the two studies. On the other hand, the solar and longwave gain factors of cloud feedback obtained from the models are not necessarily small and they are significantly different between the two studies. Although the magnitudes of the cloud gain factors are different between the different models, solar and longwave gain factors tend to compensate each other in all three models. The large inter-model difference in the seasonal variation of solar and longwave components of cloud radiative forcing was noted earlier in the analysis conduced by Cess et al.

Subtracting the gain factors of the cloud feedback from the gain factor of the overall feedback effect, we computed, for the three models and the ERBE observations, the gain factors of the feedback without the cloud feedback effect, and illustrated them in Fig. This figure indicates that the differences in solar and terrestrial gain factors among the three models are reduced substantially in agreement with the ERBE observations, when the contribution of the cloud feedback effect is excluded. In other words, the cloud feedback appears to be mainly responsible for the unrealistically large differences between the solar and terrestrial gain factors obtained from the three models. In short, solar and terrestrial gain factors obtained from the models are similar and realistic without the cloud feedback. Obviously, this does not necessarily imply that the solar and terrestrial gain factors of an individual feedback other than the cloud feedback are realistic. To confirm that they are, it is necessary to confirm that the solar and terrestrial gain factors of each simulated feedback are realistic, when they are compared with observation.

Gain factors from the ERBE observation and the three models. a The gain factor of the overall feedback, and its solar and longwave components. b The gain factor of the overall feedback minus the cloud feedback, and its solar and terrestrial components. c Gain factor of the cloud feedback. Black, dark grey, and light grey bars indicate gain factors for total radiation, solar radiation, and terrestrial radiation, respectively. The line segments attached to these bars indicate the standard error of gain factors. They are converted from the standard errors of the slope of regression between radiative flux (at the top of the atmosphere) and the global mean surface temperature, referring to Eqs.

$$f = f_{\mathrm{L}} + f_{\mathrm{S}}, \ f_{\mathrm{L}} = -\frac{1}{\lambda_0} \frac{dL^{\mathrm{FB}}}{d\bar{T}s}, \text{ and } f_{\mathrm{S}} = -\frac{1}{\lambda_0} \frac{d\overline{[S_r]}^A}{d\bar{T}s}$$

In all atmospheric models submitted to AMIP-I, an identical distribution of sea ice was prescribed, although the assigned value of albedo may differ from one model to another. Obviously, the solar gain factor obtained here is essentially determined by the prescription of sea ice. Our analysis of the simulated solar feedback would have been more meaningful if it is applied to a coupled ocean–atmosphere model, in which sea ice as well as snow cover are predicted rather than prescribed.

Contribution of Clouds to Earth's Greenhouse Effect

Clouds are vehicles for energy. They carry solar energy from the warm tropics to other parts of the globe through weather systems. But they also act as gatekeepers between

Earth and space, helping regulate the global temperature by capturing and releasing infrared (thermal) energy in the atmosphere. In this respect, clouds are like greenhouse gases. Global Warming, if clouds change as a result of global warming, the change could cause additional warming.

This image, acquired by the GOES satellite on May 30, 2010, shows thermal energy in the Western Hemisphere. The areas that are warmest and therefore emitting the most thermal energy are white and pale gray. The desert lining the Pacific coast of South America is a bright white strip in the lower center of the globe. The coldest regions emitting the least amount of thermal energy are dark gray and black. These dark spots on the globe are high clouds.

Clouds emit energy in proportion to their temperature. Low, warm clouds emit more thermal energy than high, cold clouds. This image illustrates that low clouds emit about the same amount of thermal energy as Earth's surface does. This is most clearly seen over the Pacific Ocean. The water is nearly white, while the low marine clouds are pale gray, only slightly cooler. This means that a world without low clouds loses about the same amount of energy to space as a world with low clouds.

High clouds are much colder than low clouds and the surface. They radiate less energy to space than low clouds do. The high clouds in this image are radiating significantly less thermal energy than anything else in the image. Because high clouds absorb energy so efficiently, they have the potential to raise global temperatures. In a world with high clouds, much of the energy that would otherwise escape to space is captured in the atmosphere. High clouds make the world a warmer place. If more high clouds were to form, more heat energy radiating from the surface and lower atmosphere toward space would be trapped in the atmosphere, and Earth's average surface temperature would climb.

Clouds impact temperatures in other ways as well. They also reflect energy, shading and cooling the Earth. On balance, scientists aren't entirely sure what effect clouds will have on global warming. Most climate models predict that clouds will amplify global warming slightly. Some observations of clouds support model predictions, but direct observational evidence is still limited. Clouds remain the biggest source of uncertainty (apart from human decisions to control greenhouse gas emissions) in predicting how much global temperatures will change.

Impact of the Greenhouse Effect

Climatologist believe that increasing atmospheric concentration of carbon dioxide and other "greenhouse gasses" released by human activities, such as burning of fossil fuels and deforestation, are warming the Earth. The mechanism is commonly known as the "greenhouse effect" is what makes the Earth habitable. These gasses in the atmosphere act like the glass of a greenhouse, letting the sunlight in and preventing heat from escaping. But the human activities have altered the chemical composition of the atmosphere through the buildup of greenhouse gases-primarily carbon dioxide, methane, and nitrous oxide.

Rise in environmental temperature and changes in related processes are directly connected to increasing anthropogenic greenhouse gas (GHG) emissions in the atmosphere. This rise in temperature was vehemently argued to be generally triggered by the emission of carbon based compound from fossil fuels consumption for power generation. The concentrations of carbon dioxide, methane, and nitrous oxide are all known to be increasing and in recent year, so their greenhouse gases, principally chlorofluorocarbons (CFCs), have been added in significant quantifies to the atmosphere.

History of Greenhouse Gases

The existence of the greenhouse effect was argued for by Joseph Fourier in 1824. The argument and the evidence was further strengthened by Claude Pouillet in 1827 and 1838, and reasoned from experimental observations by John Tyndall in 1859. The effect was more fully quantified by Svante Arrhenius in 1896. However, the term "greenhouse" wasn't used to describe the effect by any of these scientists; the term was first used in this way by Nils Gustaf Ekholm in 1901. In 1917 Alexander Graham Bell wrote, "the unchecked burning of fossil fuels would have a sort of greenhouse effect", and "The net result is the greenhouse becomes a sort of hot-house." Bell went on to also advocate the use of alternate energy sources, such as solar energy.

Sources of Greenhouse Gases

The most abundant greenhouse gases in Earth's atmosphere are:

- Water vapor (H_2O),
- Carbon dioxide (CO_2),
- Methane (CH_4),
- Nitrous oxide (N_2O),
- Ozone (O_3),
- Chlorofluorocarbons (CFCs).

Atmospheric concentrations of greenhouse gases are determined by the balance between sources (emissions of the gas from human activities and natural systems) and sinks (the removal of the gas from the atmosphere by conversion to a different chemical compound).The proportion of an emission remaining in the atmosphere after a specified time is the "airborne fraction" (AF). More precisely, the annual AF is the ratio of the atmospheric increase in a given year to that year's total emissions. For CO_2 the AF over the last 50 years has been increasing at 0.25 ± 0.21%/year.

By their percentage contribution to the greenhouse effect on Earth the four major gases are: water vapor, 36–70% carbon dioxide, 9–26% methane, 4–9%ozone, 3–7%. It is not physically realistic to assign a specific percentage to each gas because the absorption and emission bands of the gases overlap (hence the ranges given above). The major nonages contributors to the Earth's greenhouse effect, clouds, also absorbs and emit infrared radiation and thus have an effect on radioactive properties of the atmosphere.

In studies of the net greenhouse effect of farming systems, not only are CO_2 and CH_4 emissions important, but, due to their high specific greenhouse potential, also the site- and management-related N_{20} emissions. Model approaches have been elaborated for emission inventories on the farm level, which consider all relevant outputs; however, on the basis of partly simplified model algorithms. An overall view of the net greenhouse effect of farming systems must take into account, beside the biological C fluxes, also technical C fluxes, i.e., all CO_2 emissions involved by the input of fossil energy.

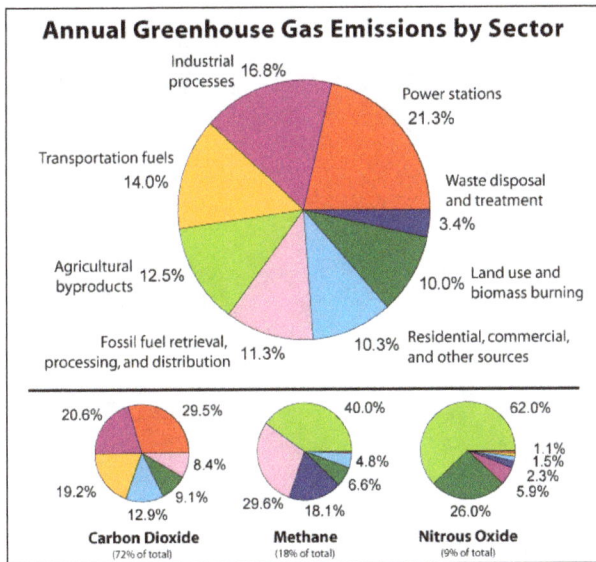

Emission of greenhouse gases.

CH_4 emissions: Methane is emitted from the production and transport of coal, natural gas, and oil. Methane emission also takes place from the decomposition of organic wastes in agriculture, in municipal solid waste, landfills and the raising of livestock. The metabolic methane emissions from livestock keeping were estimated with regard to animal species, performance and feeding. On the basis of the feed gross energy, methane releases were estimated by means of conversion factors. For quantifying the methane release from organic fertilizer during storage, the excreta output (quantity, chemical components, degradability) was chosen as the basis for calculating the methane formation potential; the amount of produced methane was then determined with regard to the storage system.

N_2O emission: N_2O emissions were estimated during agriculture and industrial activities, as well as during combustion of solid waste and fossil fuels. It was assumed, albeit very simplified, that 1.25% of the nitrogen supplied to the soils by organic and mineral fertilization, N_2 fixation and N deposition is emitted in the form of N_{2O}–N. Alternatively, a N_2O–N emission factor of 2.53% of the total N input as obtained in numerous measurements at the experimental farm was applied. The indirect N_2O emissions from gaseous NH_3 and NO_x losses as well as from N losses via reaching were quantified using emission factors.

Carbon dioxide (CO_2): Carbon dioxide enters the atmosphere through burning fossil fuels (coal, natural gas and oil), solid waste, trees and wood products, and also as a result of certain chemical reactions (e.g., manufacture of cement). Carbon dioxide is removed from the atmosphere (or "sequestered") when it is absorbed by plants as part of the biological carbon cycle.

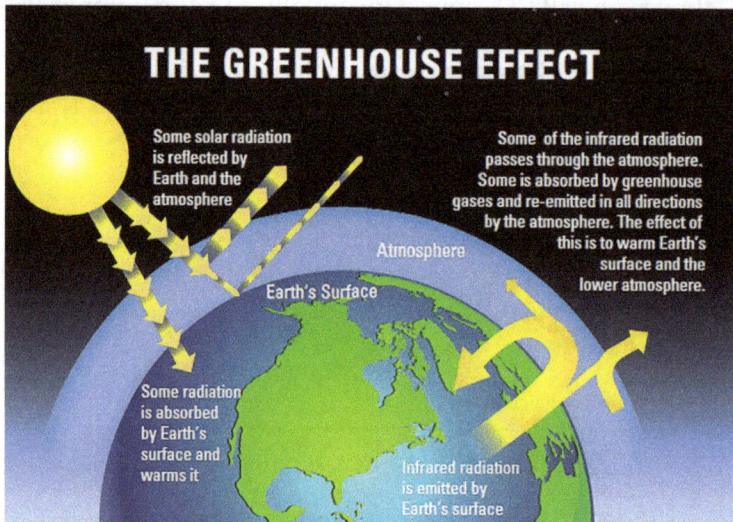

Radiation: Energy that is propagated in the form of electromagnetic waves.

- Incoming solar energy is called solar radiation.

- Solar radiation warms the earth.

- The warmed earth radiates heat. However, this is not called 'heat', but rather in scientific terms it is energy. The correct term is infrared radiation.

- The atmospheric 'blanket' is gas molecules in the atmosphere.

- Solar energy reaches the Earth's surface; the earth surface absorbs the energy and warms up.

- The warm earth surface radiates infrared radiation (IR), Greenhouse gases absorb IR leaving the surface.

- Gases are energized, then emit more radiation (IR).

- Some of this IR returns to the earth surface, warming it further.

- This process is what we call the "Greenhouse Effect".

Impact on Environment of Greenhouse Effect

Global Warming

Increase of greenhouse gases concentration causes a reduction in outgoing infrared radiation, thus the Earth's climate must change somehow to restore the balance between incoming and outgoing radiation. This "climatic change" will include a "global warming" of the Earth's surface and the lower atmosphere as warming up is the simplest way for the climate to get rid of the extra energy. However, a small rise in temperature will induce many other changes, for example, cloud cover and wind patterns. Some of these changes may act to enhance the warming (positive feedbacks), others to counteract it (negative feedbacks). Using complex climate models, the "Intergovernmental Panel on Climate Change" in their third assessment report has forecast that global mean surface temperature will rise by 1.4 °C to 5.8 °C by the end of 2100. This projection takes into account the effects of aerosols which tend to cool the climate as well as the delaying effects of the oceans which have a large thermal capacity. However, there are many uncertainties associated with this projection such as future emission rates of greenhouse gases, climate feedbacks, and the size of the ocean delay.

Sea Level Rise

If global warming takes place, sea level will rise due to two different processes. Firstly, warmer temperature cause sea level to rise due to the thermal expansion of seawater. Secondly, water from melting glaciers and the ice sheets of Greenland and the Antarctica would also add water to the ocean. It is predicted that the Earth's average sea level will rise by 0.09 to 0.88 m between 1990 and 2100.

Potential Impact on Human Life

1. Economic Impact: Over half of the human population lives within 100 kilometers of the sea. Most of this population lives in urban areas that serve as seaports. A measurable rise in sea level will have a severe economic impact on low lying coastal areas and islands, for examples, increasing the beach erosion rates along coastlines, rising sea level displacing fresh groundwater for a substantial distance inland.

2. Agricultural Impact: Experiments have shown that with higher concentrations of CO_2, plants can grow bigger and faster. However, the effect of global warming may affect the atmospheric general circulation and thus altering the global precipitation pattern as well as changing the soil moisture contents over various continents. Since it is unclear how global warming will affect climate on a regional or local scale, the probable effects on the biosphere remains uncertain.

3. Effects on Aquatic systems: The loss of coastal wetlands could certainly reduce fish populations, especially shellfish. Increased salinity in estuaries could reduce the abundance of freshwater species but could increase the presence of marine species. However, the full impact on marine species is not known.

4. Effects on Hydrological Cycle: Global precipitation is likely to increase. However, it is not known how regional rainfall patterns will change. Some regions may have more rainfall, while others may have less. Furthermore, higher temperatures would probably increase evaporation. These changes would probably create new stresses for many water management systems.

Reduction and Control Measures of Greenhouse Gases

Reduction of GHGs is central to all nations because the brunt of the problem is global and no one country or group of countries can provide its own remedy. This is why international and regional cooperation are more sought-after and have been well advocated for in the comity of global atmospheric sanity. In respect to this struggle, United Nations Framework Convention on Climate Change (UNFCCC) recently came into effect to deal with the global climate problem. This was executed in the form of international agreement comprising different countries across diverse regions to lower the dangerous concentration of anthropogenic GHGs in the atmosphere.

Clean Development Mechanism

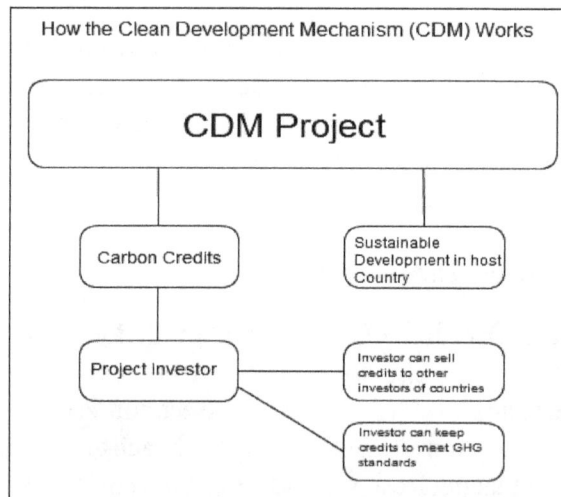

Clean Developmental Mechanism.

Clean development mechanism involve massive deployment of renewable energy technologies for power generation and carbon dioxide sequestration to promote the concept of sustainable development. Beside the GHG mitigating potential of renewable energy resources, energy security guarantee is swiftly becoming a reality with the exploitation of different renewable energy resource. Clean development mechanism is a fundamental idea of Kyoto Protocol under the canopy of the United Nations Framework on Convention on Climate Change (UNFCCC). Developing countries are more actively involved in the development of renewable power generation in line with the proposed CDM. In 2009, developing countries hosted 53% of global RE power generation. Initial idea behind the institution of CDM is to strategically lower the level of emissions due to energy generation and consumption to a sustainable intensity. However, it was envisaged that emission reduction mechanisms will be financed by the industrialized nations whereby the fund will be given to developing countries as sponsorship for renewable energy programs. After a decade and more, a good implementation result is yet to be seen and gain in the global pace of renewable power exploitation is not in line with the realistic and expected level of developments'.

Green Energy Portfolio Standard

Green energy is a type of energy produce conventionally with a reduced amount of negative environmental impact. Green energy is sometimes called renewable energy. Renewable energy application has become an essential ingredient with significant role in the expedition for GHG reduction and increasing the chance for sustainable development. Many countries have introduced and finance green energy programs to generate and consume power with minimum pollution. Green energy portfolio standard (GEPS) involves the uses of regulation to boost generation and consumption of energy from greener sources with the minimum rank of pollution propensity.

In some countries where green energy portfolio standard is strongly advocated, compulsions are placed on electric power generation companies to provide certain percentage of the national electricity demand from renewable sources as a strategic measure to lower emissions. Intergovernmental Panel on Climate Change (IPCC) direct countries to communicate their emissions from all sorts of energy related activities. Advocates of GEPS listed the benefits among which are innovation, pollution control and competition can eventually lower the per unit price of renewable power. Sustainable development of green energy can provide numerous environmental benefits alongside fossil resources conservation for far future generations.

Financing Low Carbon Energy

CO_2 emission resulting from the combustion of petroleum products contributes substantial quantity of greenhouse gas to the atmosphere. As a critical factor towards development, a secure access to modern energy is essential for development. With the current global acknowledgement on the need to reduce emissions from energy, financing low carbon energy can be used as a strategy to reduce greenhouse gas emissions. Many financing initiatives exist for funding energy projects but financing low carbon projects is indispensable especially in countries where oils are the major source of income and energy production. Driven an economy by a low-polluting energy technologies reduces the vulnerability of the human environmental sustainability. This envisioned low carbon economy can be harnessed by unlocking the untapped renewable energy resources potential. Optimization of renewable sources for energy application provides noteworthy opportunities to spread out and upgrade the energy infrastructure especially in the rural communities due to their diverseness. Via this strategic measure, the solution to energy poverty in developing regions can be provided by decentralization of the renewable energy systems. In some countries, emissions trading scheme (ETS) through carbon taxation is already implemented to control and monitor emissions.

How to reduce greenhouse effect:

- Energy conservation;

- Rising the cost of fuels;

- Developing new energy production;

- Forest protection/ Reforestation;

- Recovery of methane from garbage;

- Banning of CFC production;

- International conferences;

- National Standards of pollutants;

- Anti-pollution measures.

Prevention of Greenhouse Effect

As we burn fossil fuels such as petroleum gas and coal, carbon dioxide and various other gases get released into our environment. Heat is trapped close to the earth because of these emissions which causes what is referred to as the 'greenhouse effect'. As you probably know, greenhouse gasses can lead to significant environmental issues. The rising temperature of the Earth leads to extreme storms, higher sea levels and other issues that result from a changing climate. By working together to conserve energy, driving less and creating less waste, we will be able to reduce the carbon footprint we produce and help fight against global warming. Below are some steps to do this.

Determine the Size of your Carbon Footprint

Your carbon footprint is how much carbon you are responsible for releasing into our atmosphere due to your daily habits. Your carbon footprint becomes bigger as you burn more fossil fuels. For instance, a person who rides their bike to and from work each day will have a smaller footprint than a person who drives. You can use a free carbon footprint calculator to determine your own footprint. Have in mind that spending habits, driving habits, diet as well as other factors will be taken into account for calculating your share of carbon that is released into the environment.

Use Less AC and Heat

Caulking or installing weather stripping around windows and doors and adding insulation to your walls can reduce your costs of heating and cooling by over 25 percent since it reduces the energy you require to heat or cool your house. While sleeping or gone for the day, turn down the AC or heat. Keep moderate temperatures at all times and install a thermostat that is programmable and set it a couple degrees lower in the wintertime and higher in the summer. This can actually reduce the carbon dioxide contribution each year by 2000 pounds.

Transportation

Around 28 percent of the total greenhouse gas emissions in the United States are due to transportation. You can decrease your own personal impact by driving less, carpooling, biking, walking, using public transportation instead of cars or buying a vehicle that is more efficient. Drive slower since fuel economy becomes substantially worse when driving above 60 miles per hour. Avoid traveling by air if possible or buy carbon offsets for your flights. Braking and accelerating excessively also reduces efficiency. Don't carry excessive weight in your vehicle trunk, ensure your car is inspected and tuned up regularly and keep your tires inflated. Needless to say, this will also have a positive impact on your health.

Plant a Tree

Trees are not only meant to provide paper for you and companies like. They also help the environment. Dig and plant a tree. Trees give off oxygen and absorb carbon dioxide. Just one tree can absorb around a ton of carbon dioxide in its lifetime. In future, such policies and practices will be crucial for reducing greenhouse gas emission and stopping global climate change.

Using Renewable Energy

In the last few decades, renewables were in the main focus of every national and state government. World's brightest minds search for ways to generate electricity through solar and wind power. Such natural production of energy has major benefits for our environment. As we go forward traditional sources such as coal and oil will slowly become obsolete which is another reason why we should turn to renewable energy.

References

- Greenhouse-gas, science: britannica.com, Retrieved 11 May, 2019

- Greenhouse-Effect, Greenhouse-Gases-and-Their-Impact-on-Global-Warming: researchgate.net, Retrieved 20 August, 2019

- Greenhouseeffectcauses: conserve-energy-future.com, Retrieved 2 February, 2019

- Radiative-forcing, science: britannica.com, Retrieved 14 June, 2019

- Radiative-damping-of-annual-variation, global-mean-surface-temperature, Comparison-between-observed-and-simulated-feedback: researchgate.net, Retrieved 23 March, 2019

- Greenhouse-Effect-and-Its-Impacts-on-Environment: researchgate.net, Retrieved 21 January, 2019

- Ways-to-reduce-greenhouse-gas-emissions: interestingengineering.com, Retrieved 20 July, 2019

Chapter 5

Managing Global Warming

There are various ways to control global warming such as using alternate energy sources and by using geoengineering. The topics elaborated in this chapter will help in gaining a better perspective about the different ways to manage global warming as well as related concepts such as the cause of global warming and global cooling.

The Root Cause of Global Warming

It took more than 20 years to broadly accept that mankind is causing global warming with the emission of greenhouse gases. The drastic increase in the emission of CO_2 (carbon dioxide) within the last 30 years caused by burning fossil fuels has been identified as the major reason for the change of temperature in the atmosphere.

More than 80% of the world-wide energy demand is currently supplied by the fossil fuels coal, oil or gas. It will be impossible to find alternative sources, which could replace fossil fuels in the short or medium term. The energy demand is simply too high.

Another issue is the non-renewable characteristic of fossil fuels: It took nature millions of years to generate these resources, however we will have used them up within the next decades. Alone the shrinking supply will not make it possible to continue as usual for a longer time.

The main cause of global warming is our treatment of Nature:

- Why have warnings about climate change been ignored for more than 20 years?

- Why were ever more scientific evidence demanded to find the coherence of man-made CO_2 emissions as cause of global warming? Why wasn't common sense reason enough to act?

- Why can one still today find people who stick their head in the sand and don't want to understand what's going on in the earth's atmosphere?

- Why do most people refuse to change their personal behavior voluntary in order to reduce CO_2 emissions caused by their activities?

The answer to all these questions is a rather simple one:

In our technology and scientific minded world, we seem to have forgotten that mankind is only a relatively minor part of Nature. We ignore being part of a larger whole.

We believe to be able to control Nature instead of trying to arrange ourselves with Nature. This haughtiness is the true main cause of global warming. As a matter of fact, some people still believe that technical solutions alone would be sufficient to fight global warming.

Although we are guests on Earth, we behave as if no further visitors would arrive after us. It's like having a wild party where we destroy beds, the kitchen as well as the living room of a hotel without ever thinking about our future staying in the hotel nor about other guests arriving later.

In addition, our unit of measure is more and more often money only. What has no price tag, seems to have no value to us any more. In doing so we mix up economic growth with general well-being and financial income with personal happiness, respectively.

There is a loss of value behind this attitudes. We got blind for the true reason of our incarnation on earth. We live here to train those traits, which will finally lead to perpetual harmony with ourselves and with our environment as well as to inner calm and peace.

The ultimate global warming solutions is to behave as part of a larger whole. Many people between 20 and 65 years seem to live for the one and only purpose of earning as much money as possible in order to be able to buy as many things as possible. In this light, it is not surprising that discussions about potential solutions to fight global warming concentrate on technical measures instead of a fundamental change of our attitude to life in general and to Nature in particular.

Global Cooling

Climate alarmists constantly warn us that man-made global warming is making our world less habitable and that climate doomsday is fast approaching. But a closer look at our climate reveals a surprising climate discovery that our mainstream media have conveniently ignored for decades: the role of the sun in determining Earth's climate.

For the first time in humanity's history, our leaders could be actively devising policies — based on their defiant and biased obsession with global warming — that will render us highly vulnerable to even the slightest cooling in our climatic system.

"We are causing irreversible damage to our environment," "We are headed for a climate doomsday due to excessive warming," "Climate change may wipe out humanity" — these are our everyday news headlines. The repeated, one-dimensional doomsday cry about carbon dioxide's role in global temperature blinds the public to other causes.

CO_2 is just one of many factors that influence global temperatures. Its role in recent warming is far from dominant. Indeed, there is poor correlation between CO_2 emissions and global temperature. Between 2000 and 2018, global temperature showed no significant increase despite a steep increase in carbon dioxide emissions from anthropogenic sources. The same was the case between the years 1940 and 1970. When carbon dioxide concentration increases at a constant and steady rate and temperature doesn't follow the pattern, we can be certain that carbon dioxide is not the primary driver of global temperature.

If not CO_2, what?

Life on Earth is possible because of Earth's perfect positioning in the solar system: not too close to the sun and not too far. For centuries, academicians have acknowledged this, and climate scientists today know that the sun is the biggest influencer and driver of global temperature.

With the advent of dangerous man-made global warming theory, CO_2 has taken the limelight, and the sun has been relegated to a mere spectator. This could be warming-obsessed alarmists' biggest mistake ever.

In central Europe, for example, temperature changes since 1990 coincided more with the changes in solar activity than with atmospheric CO_2 concentration. The same has been true globally, and across centuries.

The Maunder Minimum and Dalton Minimum — periods of low solar activity — were responsible for the coldest periods of the Little Ice Age. England's River Thames froze. Whole civilizations collapsed as people starved because cold-induced poor harvests led to malnutrition that made people too weak to resist disease. Likewise, increased solar activity in the Roman Warm Period and Medieval Warm Period rought warmer temperatures on Earth, and thriving crops led to greater nutrition and lower mortality rates. Hundreds of peer-reviewed scientific papers affirm the overwhelming impact of solar activity on Earth's temperature.

Methods and Means

Demand Side Management

Climate science models rely on current and historic data of atmospheric carbon in the form of carbon dioxide, methane and chlorofluorocarbons. Carbon dioxide has long been the proxy for the determination of future anthropogenic climate changes and has been tracked continuously at the Mauna Loa Atmospheric Research Observatory since the late 1950's. Current climate models vary in projections of predicted surface-air temperature increases. Recently updated climate models from MIT indicate that the earth's surface temperature will increase by a median value of 5° Celsius by 2095 with projections as high as 7° Celsius. However, this model assumes a reduction in anthropogenic methane emissions from previous models by 300%, and disregards the large-scale increases of methane contributions from natural sources that recent field research has revealed. Even with this incorrect assumption, the model has determined that the global effects of climate change will be catastrophic.

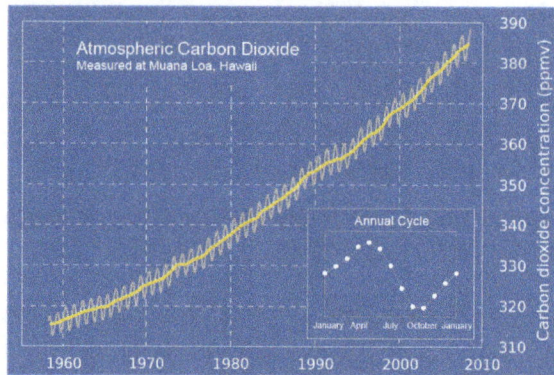

The Keeling Curve.

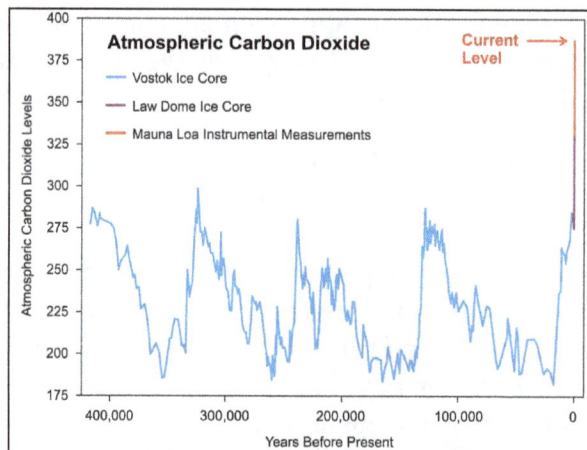

Long-Term Historic Atmospheric Carbon Dioxide.

Long-term historic trends of atmospheric carbon dioxide levels are determined by the analysis of air pockets trapped in Antarctic ice core samples taken from the Russian Vostok Antarctic Research Station and the European Project for Ice Coring in Antarctica (EPICA) Dome "C" locations in Eastern Antarctica. Long-term rainfall effects studies that used older models with a relatively gentle projected temperature increase of 3° Celsius over the next century have produced dire predictions of global drought and mass migrations from affected areas. A 2006 study indicated that regions of the globe that will experience extreme drought conditions, as measured by the Palmer Drought Severity Index (PDSI), will increase from the current value of 1% of the Earth's surface to 30% by the year 2100. Effectively, these areas will become deserts and will be unable to support sustained human activity. Recent projections show that the number of people affected by perennial water shortages in the developing world will increase from 150 million in 2000 to 1.1 billion people by the year 2050.

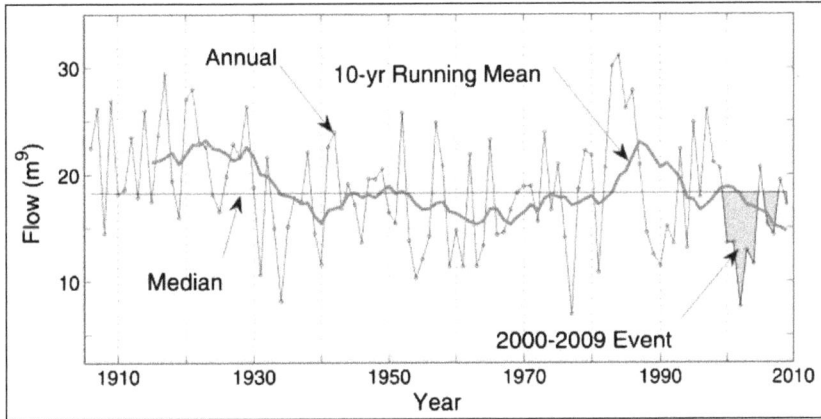

Long Term Modeling of Global Drought Patterns through the Year 2100.

The U.S. will experience areas of significantly decreased rainfall. This process will mirror the global trend of increased desertification forcing large-scale human population migrations. While temperatures will climb but still remain bearable, the decreasing ability to produce food and drinking water in regions that are severely impacted by droughts will necessitate large movements of populations from cities in the south and southwest into northern areas of the US. Projections are that the process of desertification will spread throughout the Southwest region by 2100 and that this process has already begun with the recent 10-year drought happening at this time.

Carbon dioxide in the Earth's atmosphere does not remain indefinitely. Natural forces work to remove carbon dioxide gas from the atmosphere on a regular basis. The global warming potential lifetime of a specific "pulse" of injected carbon dioxide in the Earth's atmosphere is measured to be approximately 100 years. However, the Earth's ability to remove carbon dioxide is greatly affected by air and sea surface temperatures. It has recently been determined that the projection of a 5° Celsius increase of surface air temperatures will lead to a decline in the natural removal rates of carbon dioxide by 301 Giga-tons per year. This correlates to a total weight of 82 Giga-tons of carbon. In comparison, the total annual weight of carbon, in the form of carbon dioxide, removed through the process of land-based photosynthesis is only 60 Giga-tons per year.

Current U.S. national security estimates agree that 2030 is the approximate timing for ice-free summer months in the Arctic Ocean.

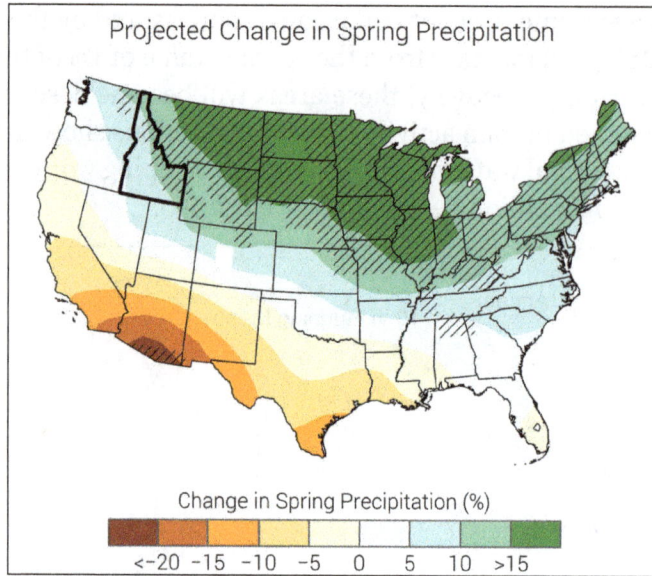

Projections of Decreases in Spring Rainfall Totals in the South western United States.

Sudden Climate Change (Methane)

Indicated long-term (100 year) effects of the carbon content of the atmosphere. New information indicates a rapid speed-up of climate change due to methane. Methane now has a very high potential for causing a short range (20 year) climate change feedback. The possibility is for an abrupt (decadal) severe climate change. On a mass-to-mass direct comparison with carbon dioxide, methane has up to a 105-fold greater warming potential for a 20-year time horizon. It also has a 33-fold greater warming potential for a 100-year time horizon when compared to carbon dioxide. Current estimates indicate that the increase of global atmospheric methane since pre-industrial times is a significant contributor to the currently observed anthropogenic climate change:

> "The global atmospheric methane burden has more than doubled since preindustrial times and this increase is responsible for about 20% of the estimated change in direct radiative forcing due to anthropogenic greenhouse-gas emissions."

As arctic sea ice levels retreat, very recent observational data from deep-ocean buoys has determined that deep-arctic ocean temperatures are rapidly rising. A draft study of waters off the coast of northwest Greenland has found temperature rises at a depth of 350 meters as great as 0.63 °C per decade and attributes this to a "changing ice regime". Significant amounts of methane are currently sequestered within sediments located underneath the relatively shallow Eastern Siberian Arctic Shelf. As sea temperatures rise in this location, the methane clathrates13 contained within the sub-sea sediments are rapidly melting and releasing large quantities of stored methane. If only 1% of this area's sub-sea methane is released into the atmosphere it will increase the earth's current methane burden by 300 to 400 percent. Since the doubling of the atmospheric methane burden between 1750 and now currently contributes 20% to the observed effects of anthropogenic climate change, this release would result in an increase in the radiative forcing effects of anthropogenic greenhouse gasses in the earth's atmosphere by 100% to 140% in a very short amount of time. In 2008 The U.S. Department of Energy recognized this release as a credible threat to national security and established the Investigation of the Magnitudes and Probabilities of Abrupt Climate Transitions (IMPACTS) Project, a collaboration of six national energy labs working to determine the potential for abrupt climate changes to occur.

"Plausible scenarios could lead to methane becoming more important than CO_2 as a greenhouse gas on a time-scale of decades, with the associated warming leading to further hydrate dissociation, as well as terrestrial permafrost melting, that will release additional methane and be self-sustaining." - Investigation of the Magnitudes and Probabilities of Abrupt Climate Transitions (IMPACTS) Project.

Barriers to Thought and Action

Next, we look at barriers that make response to climate change difficult. These include valuing inappropriate conventions of economics and finance over physical reality, the weakening of the middle and professional classes, and the promotion of anti-science doubt.

First, with the exception of ecological economics, economics and finance (whether capitalist or Marxist) are very weak in treatment of physical limits. In their conventional ways of looking at the world, the costs to all of us from corporate production processes are obfuscated. Current economics can only be corrected by situating it within physics. With this shift, the interlinked problems of overpopulation, pollution, global warming, and increasing energy resource scarcity become the dominant features of economic reality. The conventional focus on profits and returns, financial discounting of the future and on the "the market" as a solution is outdated and misdirected. Without this correction, economics offers inaccurate models of production without resource depletion, without pollution and without the warming (except as unimportant residuals). It typically indicates that a solution to problems is to "grow our way out". Conventional financial perspectives are based on mechanical models that discount the future.

Second, we have the weakening of the middle and professional classes since 1970. The attentive public on important issues in a democracy is made up of persons who have the time, security and resources to follow issues beyond their immediate needs. Currently, as one DSM Manager has put it, we are trying to do DSM in a consumer ridden, stratified culture. In this culture, radically increasing income inequalities are a huge barrier to market driven programs. For example, some home audit programs now try to take income into account so that well-to-do families are provided with different measure recommendations than low income families, a process that is not rational from a physical perspective. The gradual defunding of the middle, upper middle, and portions of the upper classes in the United States began in about 1970 with the continuing transfer of wealth and income to the extreme upper 2% of US households. This trend weakens the ability of market-based DSM. It also weakens government action by undermining the tax base necessary to provide a strong collective response to DSM and global warming. The warming, interacting with the intentional "creative destruction" endemic to a corporate market economy and focus on growth of the worldwide profit system pushes these tensions further. As weather becomes "unnatural," as inflation grows, as food production becomes more difficult, as the job structures start modestly to recover and then fail again, it will become harder to do effective DSM. This is especially true as portions of the social safety net are retracted for the poor and middle class, as pay is cut for professional workers; and as government employees are dismissed from essential positions. For example, we are having some very good solar promotion with good take up at the local level but the scale of participation is limited by the draining of economic capability with sustained 22%-25% unemployment. By contrast, what we need for either market-based or collectively financed DSM is a vibrant middle, upper middle and upper class that can provide a basis for a pro-science government policy as it affects both resource depletion and global warming.

Third, there is intentional creation of doubt. From the 1920's some critical industries have worked directly to create doubt on science issues through various strategies: directly, through think tanks, and through capture of federal regulatory agencies. From the 1950's, a relatively few individuals and industry funded think tanks have created doubt about tobacco and cancer, second-hand smoke, the hole in the ozone layer, the need to control DDT, the need to control acid rain, and (now) the need to mobilize to control global warming. The current major political effort to weaken the Environmental Protection Administration (EPA) is supported by the work of corporately funded "free market" think tanks. Science recognizes resource depletion, pollution, and global warming at the *center* of material reality. Some corporations directly affected (and others on simple economic model or viewpoint grounds) would prefer that the limits implied by reality did not exist because they interfere with their "liberty" and "certain kinds of liberty are not sustainable—like the liberty to pollute". Once scientific thinking asserts priority over conventions of economics, finance, and production it is a threat to the basis of a capitalist form of organizing the economic system.

"For the sake of a cleaner planet, should Americans wear dirtier clothes?" Tierney highlights the economist's "rebound effect" or "Jevons Paradox," named after an 1850's economist who observed that increased efficiency of steam engines led to increasing use of steam engines and to more coal consumption. Modern economists Jenkins, Nordhaus and Shellenberger are quoted as suggesting that increasing energy efficiency in a steel plant in China is likely to increase production. Tsao et al are cited as showing people spend about the same percentage of income on light today (LEDs, etc.) as they did in 1700 (candles, etc.). Further, Saunders, a co-author with Tsao is quoted as saying that LEDs and other advances in energy efficiency do not reduce energy use: "We find the opposite is true." The National Research Council is cited as authority for a finding of additional highway deaths each year due to cars being less safe because they are more efficient. Tierney concludes, quoting Sam Kazman of the Competitive Enterprise Institute ("a free-market-oriented nonprofit research group"): "Efficiency mandates results have ranged from costly fiascos, such as once-dependable top-loading washers that no longer wash, to higher fatalities in cars downsized by fuel-efficiency rules. If the technologies were so good, they wouldn't need to be imposed on us by law." What this topic is really about is not science but the liberty to choose in a so-called "free market" to avoid the collective social responsibility embodied in energy efficiency programs and regulation to improve product efficiency and reduce greenhouse gas emissions. It is about a mechanical model.

Organizational Response

The DSM community is responding with strong improvements in building codes, the movement to near zero energy buildings, the new social rationing technologies inherent in the smart grid and smart meters and, for example, the European Committee for an Energy Efficient Economy's development of the cost optimality project selection criterion. Many other projects are underway including "deep" DSM research by joint utilities in Massachusetts and by a coalition put together by Affordable Comfort.

Part of this response involves the question of agency. By "agency" we mean the organizational structure through which DSM will be accomplished. We will contrast two extremes, current organization and total social mobilization. We expect DSM to slowly transition from the first to the second. The first case is what we have now. The second is based on the precedent of the WWII Office of Price Stabilization. Currently, DSM is administered by the states and provinces of North America as independent jurisdictions and regulated by Utility Boards or Public Utility Commissions. However, certain programs such as Energy Star and Low-Income Weatherization are federal programs though administered through the jurisdictions. Complementing this formal structure, much of what makes DSM work are the regional administrative and coordinating groups (NEEP, NEEA MEEA, SEEA), professional groups (Association of Energy Engineers, American Energy Services Professionals), working groups on buildings and renewable energy at the national laboratories and the American Council for an Energy Efficient Economy. Most of the DSM achievements are accomplished by private sector program delivery agents and results are primarily assessed by a network of private sector evaluation, measurement and verification firms and individual consultants. This kind of porous and multi-centered structure is excellent for innovation for a climate change framework. And, as shown by the many recent innovations in DSM cost tests, is innovating in the general direction of movement towards creating the capabilities to introduce deep DSM suited to our climate change problems.

Moving Towards an Adequate Cost Test

There are three frameworks for structuring Demand-Side Management efforts. These are resource acquisition, market transformation, and climate change. Formal Demand-Side Management (DSM) began as a component of utility Integrated Resource Planning with the resource acquisition framework. The market transformation framework further incorporates market shifts that permit the cost per unit of a transformed product to drop to virtually zero in the long-term. The high cost units of the first or middle years, when melded with the zero cost units of the late years show an average cost competitive with the marginal plant or the market price of the alternative generation mix. The climate change framework requires another transformative change of this kind.

The climate change framework has not been fully worked out, but there are many people developing 'work-arounds' that provide indications of this framework coming together. Wisconsin lowered the discount rate of its Total Resource Cost Test (TRC) to a low social discount rate, several states call their primary cost test the TRC but have introduced small changes that make the test more like a societal test (including some non-energy benefits and/or an environmental adder), some states move the application of the TRC to the program or portfolio level. California is working on a comprehensive Integrated Demand Side Management (IDSM) test that will include valuation of energy embodied in water and several other factors.

We outline a much simpler approach as a trial climate change framework for testing climate change DSM programs. This framework is rooted in the second law of thermodynamics (heat energy) in its application on a human scale and takes into account this law of physics as the primary economic law for the framework. The theory of the climate framework for DSM draws on the environmental and climate sciences.

In the climate approach, in contrast to earlier DSM cost-testing, policy goals for energy efficiency must be set to accomplish goals by specific dates. This necessarily means that the future is not discounted – it is either counted as equal to the present or, preferably, counted as more important than the present. At the same time, the appropriate primary cost test shifts to a combination of the Administrator's Cost Test and the BTU/dollar test. In addition, it will be necessary to include within the program authorization procedure a review by a small team of engineers and DSM policy people. In itself, the Program Administrator's Cost Test frees the field for climate change programs by explicitly treating cost sharing as leverage. The BTU/Dollar test does not discount energy so it makes it possible to fund an incentive for a Passivehaus designed to last 150 years and to compare possible projects using their actual energy streams (1 year for one, 17 for another, and 150 for another). Note that by not discounting the future, the value of conserved energy does not go to zero at approximately 17 years as an artifact of the cost-testing method. Until the DSM climate framework becomes established it is conservative and prudent to include an expert team review as part of the process of project authorization. This is to prevent "green folly" projects.

To extend the climate perspective further, we need to use the BTU/Dollar test simply as a tool for optimizing packages of energy efficiency improvements. Beyond this test, we need to be able to specify homes, business, industrial, and institutional building characteristics that will last 150 years under the specific local and regional climate conditions of the new world that we face. If this means walls three feet thick, then walls have to be three feet thick regardless of cost. Physics rather than economics or finance has to dictate the new built environment. This is a very different perspective than the perspective that gave rise to the current cost tests and the new tests that are being evolved.

To push perspective beyond this point, energy efficiency is only one part of the solution to climate change. It is essential, but only a component. We envision that a command and control economic system will be necessary to deploy both energy efficiency and direct conservation (enforced rationing and voluntary and involuntary modifications of lifestyles to "do without"). What we face is a sustained emergency longer than any other that humans have ever consciously faced and it is likely that even if we win only a small fraction of the current world population will make it through.

Total Social Mobilization

In the face of catastrophic global climate change, energy efficiency must take on, as President Carter said long ago, "the moral equivalent of war." It will have to be deployed as a set of coordinated global warming DSM strategies and programs. Far from free market economics and as feared by free market anti-science interests, DSM will have to be deployed on a command and control basis. Total social mobilization means central federal control of global warming effort, including DSM strategies and programs operated through the jurisdictions. By analogy, the theory and practice of this model is best worked out by Galbraith in the operations of the World War II Office of Price Control (OPC). During WWII it was necessary for all combatant nations to operate their economies outside of the market system. In the US, the OPC allocated resources by command and control to the war industries to meet the enormous material needs of the fighting forces. Galbraith shows that for such a system (which breaks most of the rules and expectations of market economics) to work, both prices and wages throughout the economy had to also be controlled. And, for control of production allocations to work, a successful rationing of food, medical, and all consumer goods was necessary. This system worked well throughout World War II and was dismantled after victory.

Through 2020

Ocean levels rise, CO_2 level rises to about 430 ppm, tundra and undersea ice continue to rapidly outgas methane. Movements of small numbers of climate refugees has already begun and will become large. In some lucky areas, planned relocation will be made with the support of national governments or the UN. In others, people will just start moving. Storms will become more severe. Crop yields will slowly decline and farmers will experience more frequent "freak weather."

DSM continues much as it is now. With many independent centers in different political jurisdictions, working groups and associations there is a high potential for innovation. In particular, the old DSM Total Resource Cost Test (TRC) will be continually challenged, both in the ways suggested in this paper and in much more complex technical approaches as in the California Integrated DSM models. The "deep" energy projects of the joint Massachusetts utilities and Affordable Comfort will produce usable results. There is a potential across jurisdictions to develop variants of a DSM Climate Change framework and to try out a few climate change DSM programs on a trial basis. Good initial candidates for climate demonstration projects and trial programs are universal transformation to LED street lighting (in cold areas remember to select a brand that warms the electronics), fire, police, and medical facilities, schools and military facilities. Climate change housing is also a good focus (include climate change ready permanent emergency housing for the expected masses of climate change immigrants). In Europe, the ECEEE will proceed with the continuing development of strong improvements in building codes, the movement to near zero energy buildings, and with development of the cost optimality project. In both the North America and Europe new social rationing technologies represented in the smart grid and smart meters will be operated for rationing and social control.

Through 2050

Physical forces will have moved the warming forward perhaps very rapidly based on methane releases or more slowly based on carbon releases to the atmosphere. The changes required for dealing with the extent of global warming will cause the loss of corporate and individual liberty to misuse, pollute, and destroy the world. But loss of liberty will be not be noticeably worse than in WWII when major nations were in a contest for their continued existence. For the most part, people will try to make things work. Water levels and greenhouse gas levels will continue to rise. Summer Arctic sea ice will be non-existent. The oceans will be much more acidic causing die offs of ocean life and including stocks of food fish. We will still be below a runaway greenhouse effect. Farming will have become challenging in many areas. Both plants and animal life will have been trying for some years to migrate, resulting in extensive die offs but some successes. Invasive species entering new areas will cause major changes in natural food networks and insects and diseases will likely become pestiferous in most places.

The former great farming areas of the US and Canada will be drying out yet subject to frequent intense storms, including tornados and also be subject to rapid evaporation. Farming from Oregon through British Columbia and from Maine through the Maritime Provinces will remain viable. Land will become available in Northern Canada, but not topsoil. Greenland and a part of Antarctica will start to be available for precarious settlement. There is clean topsoil from ancient jungles in the part of Antarctica that becomes available for settlement. Some governments will collapse, but other will remain intact, very small elites continue to have access to the good life, and in lucky areas nearly all institutions will continue to function though at reduced levels. The entire world society will be under strain and human population will drop substantially. Still, the changes are tolerable in the whole. In the carbon future, New Hampshire is now becoming South Carolina, but South Carolina is a good place. In the methane future, things have speeded up and we are far beyond that point.

Nations will necessarily shift to command and control to support the armed defense of territory and resources and to insure allocation of prices and wages as well as production and consumer goods. There may be resource wars which will further drain social resources and weaken governments.

DSM in this period will have moved to a climate change framework. Coal-friendly cost tests like the TRC Test will be quaint history. Energy policy will have to include new construction of nuclear plants on a production basis to attempt to make up for the close out of coal. DSM energy efficiency and renewable energy will be combined and projects will be set according to energy targets in specific locations and specific years. It is likely that the officers of AESP and ACEEE will be pulled into government and given military ranks as happened to essential civilian professionals during World War I. DSM will convert to a command and control basis for the long duration of sustained emergency.

Through 2075 or 2111

The CO_2 models suggest that it is unlikely that we will be able to prevent a runaway greenhouse effect in the long range (100 years). Looking beyond carbon as a proxy for all greenhouse gasses to specifically consider methane, these problems may speed to a twenty to thirty-five year range rather than a one-hundred year rage. With the tipping point for release of methane clathrates now passed and fracking releasing perhaps twenty times more methane than ordinary gas wells and proceeding at full speed, methane moves from a minor to a major concern. This could speed up the climate effects on the order of a decade or a few decades.

Alternative Energy Sources

The potential issues surrounding the use of fossil fuels, particularly in terms of climate change, were considered earlier than you may think. It was a Swedish scientist named Svante Arrhenius who was the first to state that the use of fossil fuel could contribute to global warming, way back in 1896.

The issue has become a hot-button topic over the course of the last few decades. Today, there is a general shift towards environmental awareness and the sources of our energy are coming under closer scrutiny. This has led to the rise of a number of alternative energy sources. While the viability of each can be argued, they all contribute something positive when compared to fossil fuels.

Lower emissions, lower fuel prices and the reduction of pollution are all advantages that the use of alternative fuels can often provide. Here we examine eleven of the most prominent alternative fuel sources and look at the benefits they offer and potential for increased uptake in the coming years.

Alternative Energy Sources

Hydrogen Gas

Unlike other forms of natural gas, hydrogen is a completely clean burning fuel. Once produced, hydrogen gas cells emit only water vapor and warm air when in use. The major issue with this form of alternative energy is that it is mostly derived from the use of natural gas and fossil fuels. As such, it could be argued that the emissions created to extract it counteract the benefits of its use.

The process of electrolysis, which is essential for the splitting of water into hydrogen and oxygen, makes this less of an issue. However, electrolysis still ranks below the previously mentioned methods for obtaining hydrogen, though research continues to make it more efficient and cost-effective.

Tidal Energy

While tidal energy uses the power of water to generate energy, much like with hydro-electric methods, its application actually has more in common with wind turbines in many cases. Though it is a fairly new technology, its potential is enormous. A report produced in the United Kingdom estimated that tidal energy could meet as much as 20% of the UK's current electricity demands.

The most common form of tidal energy generation is the use of Tidal Stream Generators. These use the kinetic energy of the ocean to power turbines, without producing the waste of fossil fuels or being as susceptible to the elements as other forms of alternative energy.

Biomass Energy

Biomass energy comes in a number of forms. Burning wood has been used for thousands of years to create heat, but more recent advancements have also seen waste, such as that in landfills, and alcohol products used for similar purposes.

Focusing on burning wood, the heat generated can be equivalent to that of a central heating system. Furthermore, the costs involved tend to be lower and the amount of carbon released by this kind of fuel falls below the amount released by fossil fuels.

However, there are a number of issues that you need to consider with these systems, especially if installed in the home. Maintenance can be a factor, plus you may need to acquire permission from a local authority to install one.

Wind Energy

This form of energy generation has become increasingly popular in recent years. It offers much the same benefits that many other alternative fuel sources do in that it makes use of a renewable source and generates no waste.

Current wind energy installations power roughly twenty million homes in the United States per year and that number is growing. Most states in the nation now have some form of wind energy set-up and investment into the technology continues to grow.

Unfortunately, this form of energy generation also presents challenges. Wind turbines restrict views and may be dangerous to some forms of wildlife.

Geothermal Power

At its most basic, geothermal power is about extracting energy from the ground around us. It is growing increasingly popular, with the sector as a whole experiencing five percent growth in 2015.

The World Bank currently estimates that around forty countries could meet most of their power demands using geothermal power. This power source has massive potential while doing little to disrupt the land. However, the heavy upfront costs of creating geothermal power plants has led to slower adoption than may have been expected for a fuel source with so much promise.

Natural Gas

Natural gas sources have been in use for a number of decades, but it is through the progression of compression techniques that it is becoming a more viable alternative energy source. In particular, it is being used in cars to reduce carbon emissions. Demand for this energy source has been increasing. In 2016, the lower 48 states of the United States reached record levels of demand and consumption.

Despite this, natural gas does come with some issues. The potential for contamination is larger than with other alternative fuel sources and natural gas still emits greenhouse gases, even if the amount is lower than with fossil fuels.

Biofuels

In contrast to biomass energy sources, biofuels make use of animal and plant life to create energy. In essence they are fuels that can be obtained from some form of organic matter.

They are renewable in cases where plants are used, as these can be regrown on a yearly basis. However, they do require dedicated machinery for extraction, which can contribute to increased emissions even if biofuels themselves don't.

Biofuels are increasingly being adopted, particularly in the United States. They accounted for approximately seven percent of transport fuel consumption as of 2012.

Wave Energy

Water again proves itself to be a valuable contributor to alternative energy fuel sources with wave energy converters. These hold an advantage over tidal energy sources because they can be placed in the ocean in various situations and locations.

Much like with tidal energy, the benefits come in the lack of waste produced. It is also more reliable than many other forms of alternative energy and has enormous potential when used properly.

Again, the cost of such systems is a major contributing factor to slow uptake. We also don't yet have enough data to find out how wave energy converters affect natural ecosystems.

Hydroelectric Energy

Hydroelectric methods actually are some of the earliest means of creating energy, though their use began to decline with the rise of fossil fuels. Despite this, they still account for approximately seven percent of the energy produced in the United States.

Hydroelectric energy carries with it a number of benefits. Not only is it a clean source of energy, which means it doesn't create pollution and the myriad issues that arise from it, but it is also a renewable energy source.

Better yet, it also offers a number of secondary benefits that are not immediately apparent. The dams used in generating hydroelectric power also contribute to flood control and irrigation techniques.

Nuclear Power

Nuclear power is amongst the most abundant forms of alternative energy. It creates a number of direct benefits in terms of emissions and efficiency, while also boosting the economy by creating jobs in plant creation and operation.

Thirteen countries relied on nuclear power to produce at least a quarter of their electricity as of 2015 and there are currently 450 plants in operation throughout the world.

The drawback is that when something goes wrong with a nuclear power plant the potential for catastrophe exists. The situations in Chernobyl and Fukushima are examples of this.

Solar Power

When most people think of alternative energy sources they tend to use solar power as an example. The technology has evolved massively over the years and is now used for large-scale energy production and power generation for single homes.

A number of countries have introduced initiatives to promote the growth of solar power. The United Kingdom's 'Feed-in Tariff' is one example, as is the United States' 'Solar Investment Tax Credit'.

This energy source is completely renewable and the costs of installation are outweighed by the money saved in energy bills from traditional suppliers. Nevertheless, solar cells are prone to deterioration over large periods of time and are not as effective in unideal weather conditions.

Geoenginnering

Climate geoengineering refers to large-scale schemes for intervention in the earth's oceans, soils and atmosphere with the aim of reducing the effects of climate change, usually temporarily.

Types of Geoengineering

The main categories of proposed geoengineering techniques are:

- Solar radiation management: SRM techniques attempt to reflect sunlight back into space, and include a range of ideas, from orbiting mirrors, tonnes of sulphates sprayed into the stratosphere, and modifying clouds, plants and ice to make them more reflect more sunlight.

- Carbon dioxide removal: These proposals posit that it's possible to suck carbon out of the atmosphere on a massive scale, using a combination of biological and mechanical methods, from seeding the ocean with iron pellets to create plankton blooms to creating forests of mechanical "artificial trees".

- Earth radiation management: ERM proponents suggest that negative effects of climate change can be offset by allowing heat to escape into space – for example, by thinning cirrus clouds.

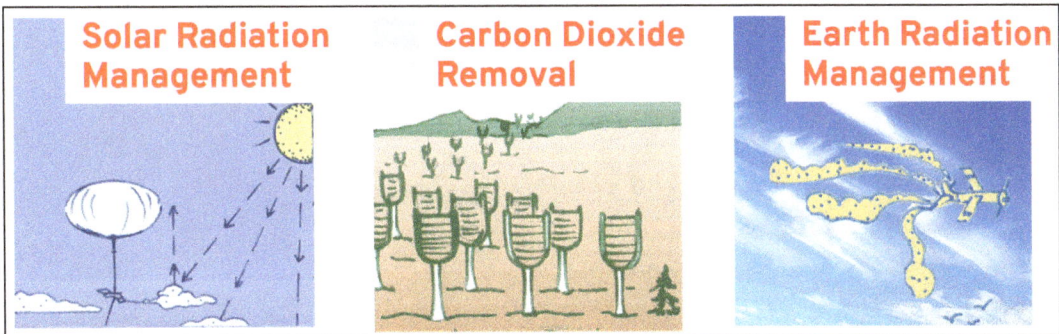

Solar Radiation Management   Carbon Dioxide Removal   Earth Radiation Management

Characteristics

Geoengineering is a false solution to the climate crisis that aims to address the symptoms of climate change but ignores and enables the root causes to continue. While there are a variety of geoengineering techniques and technologies, each with their own ecological and social implications, a few important characteristics apply to all techniques:

- They don't really exist: To date, the claims made about geoengineering techniques are purely based on speculation, and are – effectively by definition – not real technologies.

- Favoured by global north, backed by billionaires: Most of the political and financial support for geoengineering comes from a small group of elite engineers, a handful of billionaires, and a growing group of right wing politicians (many of them former climate deniers).

- Ecological impacts are huge: The sheer scale of many of of these proposals would have massive negative and unpredictable impacts on the environment – air, land and sea – which would be disproportionately borne by the global south.

Impacts and Critiques of Specific Techniques

Climate geoengineering proposals represent efforts to manipulate the climate on a global scale, but each proposed technique brings its own environmental and social impacts.

References

- Main-cause-of-global-warming-solutions: timeforchange.org, Retrieved 25 June, 2019

- Global-cooling-the-real-climate-threat: cornwallalliance.org, Retrieved 29 April, 2019

- Alternative-energy-sources: renewableresourcescoalition.org, Retrieved 9 August, 2019

- What-is-geoengineering: geoengineeringmonitor.org, Retrieved 19 May, 2019

Permissions

All chapters in this book are published with permission under the Creative Commons Attribution Share Alike License or equivalent. Every chapter published in this book has been scrutinized by our experts. Their significance has been extensively debated. The topics covered herein carry significant information for a comprehensive understanding. They may even be implemented as practical applications or may be referred to as a beginning point for further studies.

We would like to thank the editorial team for lending their expertise to make the book truly unique. They have played a crucial role in the development of this book. Without their invaluable contributions this book wouldn't have been possible. They have made vital efforts to compile up to date information on the varied aspects of this subject to make this book a valuable addition to the collection of many professionals and students.

This book was conceptualized with the vision of imparting up-to-date and integrated information in this field. To ensure the same, a matchless editorial board was set up. Every individual on the board went through rigorous rounds of assessment to prove their worth. After which they invested a large part of their time researching and compiling the most relevant data for our readers.

The editorial board has been involved in producing this book since its inception. They have spent rigorous hours researching and exploring the diverse topics which have resulted in the successful publishing of this book. They have passed on their knowledge of decades through this book. To expedite this challenging task, the publisher supported the team at every step. A small team of assistant editors was also appointed to further simplify the editing procedure and attain best results for the readers.

Apart from the editorial board, the designing team has also invested a significant amount of their time in understanding the subject and creating the most relevant covers. They scrutinized every image to scout for the most suitable representation of the subject and create an appropriate cover for the book.

The publishing team has been an ardent support to the editorial, designing and production team. Their endless efforts to recruit the best for this project, has resulted in the accomplishment of this book. They are a veteran in the field of academics and their pool of knowledge is as vast as their experience in printing. Their expertise and guidance has proved useful at every step. Their uncompromising quality standards have made this book an exceptional effort. Their encouragement from time to time has been an inspiration for everyone.

The publisher and the editorial board hope that this book will prove to be a valuable piece of knowledge for students, practitioners and scholars across the globe.

Index

www.ingramcontent.com/pod-product-compliance
Lightning Source LLC
Chambersburg PA
CBHW062002190326
41458CB00009B/2946